MEMORY

The MIT Press Essential Knowledge Series

A complete list of books in this series can be found online at
https://mitpress.mit.edu/books/series/mit-press-essential-knowledge-series.

MEMORY

FERGUS CRAIK AND LARRY JACOBY

The MIT Press | Cambridge, Massachusetts | London, England

The MIT Press would like to thank the anonymous peer reviewers who provided comments on drafts of this book. The generous work of academic experts is essential for establishing the authority and quality of our publications. We acknowledge with gratitude the contributions of these otherwise uncredited readers.

This book was set in Chaparral Pro by New Best-set Typesetters Ltd. Printed and bound in the United States of America

Library of Congress Cataloging-in-Publication Data

Names: Craik, Fergus I. M., author. | Jacoby, Larry, author.
Title: Memory / Fergus Craik and Larry Jacoby.
Description: Cambridge, Massachusetts : The MIT Press, [2023] | Series: The MIT Press essential knowledge series | Includes bibliographical references and index.
Identifiers: LCCN 2022011631 (print) | LCCN 2022011632 (ebook) | ISBN 9780262545204 (paperback) | ISBN 9780262373579 (epub) | ISBN 9780262373586 (pdf)
Subjects: LCSH: Memory.
Classification: LCC BF371 .C7523 2023 (print) | LCC BF371 (ebook) | DDC 153.1/2—dc23/eng/20220720
LC record available at https://lccn.loc.gov/2022011631
LC ebook record available at https://lccn.loc.gov/2022011632

10 9 8 7 6 5 4 3 2 1

CONTENTS

SERIES FOREWORD

The MIT Press Essential Knowledge series offers accessible, concise, beautifully produced pocket-size books on topics of current interest. Written by leading thinkers, the books in this series deliver expert overviews of subjects that range from the cultural and the historical to the scientific and the technical.

In today's era of instant information gratification, we have ready access to opinions, rationalizations, and superficial descriptions. Much harder to come by is the foundational knowledge that informs a principled understanding of the world. Essential Knowledge books fill that need. Synthesizing specialized subject matter for nonspecialists and engaging critical topics through fundamentals, each of these compact volumes offers readers a point of access to complex ideas.

Just about everyone is interested in memory. Initially this interest stems from the experience of memory's astonishing strengths and weaknesses. How can it be that the smell of baking bread transports us back sixty years to grandma's kitchen, while a minute later we can't recall the name of a well-known colleague (which in turn pops back to mind spontaneously two hours later)? We can remember details of some emotional childhood event with great clarity, yet forget where we left the car keys twenty minutes ago. We remember hearing the news of some shockingly unexpected event—the death of Princess Diana or the destruction of the World Trade Center towers on 9/11—not only hearing the news, but where we were, who was there, and what happened next. Yet studies have shown that such "crystal clear" memories are often false. How can that be when we are certain that some detail was part of the experience? What about extraordinary feats of memory—for example, the ability of chess masters to hold eight or ten games in mind simultaneously as play progresses, or a musical maestro to conduct a long, complex symphony without the aid of the score? What happens in the brain when older people become pathologically forgetful in the early stages of Alzheimer's disease? And if you start forgetting names and intended grocery purchases when you are sixty-three, is that simply "normal aging" or is it a sign of encroaching dementia? Finally, how effective are brain games and dietary supplements at improving memory, attention, and concentration?

This book attempts to answer at least some of these intriguing questions. We describe cases of extraordinary memory and the efforts made by researchers to understand them as well as cases of memory failure due to brain damage, disease, emotional trauma, and other causes. Everyday strengths and weaknesses have been investigated by cognitive psychologists in laboratory studies for more than a hundred years, and we look at some of their methods, findings, and implications for real-life remembering. Patients with memory problems have provided another major source of knowledge about how memory is represented and organized. We discuss a few famous cases, and what they reveal about the registration, retention, and retrieval of specific events and general knowledge in the brain. The biological basis of memory and learning is a further topic of great interest to researchers. In fact, much of the excitement in current memory research stems from interactions among geneticists, molecular biologists, neurophysiologists, and behavioral psychologists, often working with animals to gain clues to the nature of memory and its evolutionary history. The introduction of neuroimaging methods—brain scanning—is an important development in this research area. The techniques allow researchers to study the brain changes accompanying the cognitive processes of encoding (the registration of new information) and retrieval.

Memory can therefore be portrayed in several different ways. Our own experiences of memory—the rich recollection of a family holiday or the frustration of failing to recall a name we know well—are probably most important to us, but when scientists study memory, other aspects come into play. Cognitive psychologists document the conditions that affect retention and forgetting; neuroscientists explore the underpinnings of memory in terms of brain networks; and molecular biologists study how such networks are formed and activated during learning and remembering. These different areas of research are all crucial. A final understanding of what human memory *is* will require a demonstration of how findings from these various levels of analysis map onto each other, how changes in single neurons combine to form changes in extensive brain networks, and how differences in network organization translate to differences in behavior and to differences in our personal experience.

UNDERSTANDING MEMORY

An Evolving Story

Some Early Metaphors

In this first chapter, we review the various ways in which philosophers and scientists over the centuries have attempted to understand what memory *is*, and how it works. Initially, the descriptions were frankly metaphors, but they typically make important points about the nature of learning and memory. One obvious feature of such metaphors is that they reflect the communication technology of the day, and this is true even of early metaphors. Both Plato and Aristotle likened memory to the wax tablets used by their students to record their lessons. A signet ring pressed into the soft wax "left an impression"—a term still in use today—and the record remained until it was overwritten by a further mark or the wax melted. These ancient philosophers also suggested that differences in memory ability could be understood in terms of differences in the consistency of the material receiving impressions. If the wax was too soft and runny, it would not retain a sharp image, and if it was too hard (as perhaps in elderly people), it would not record the impression. Ingenious though they were, however, these early thinkers were unable to show what corresponded to the wax tablet in real humans.

Plato offered another metaphor illustrating the difference between possessing a specific piece of knowledge, or memory of an event, and being able to bring it to conscious awareness at a given moment. He told his students to imagine they had caught birds of different species and built a large aviary to keep them in. On entering the aviary, it might be possible to catch a bird and hold it temporarily in hand—corresponding to a remembered event in the person's mind. On release of the bird, that specific memory is no longer in conscious awareness although it is still retained in the overall system. This distinction between the many past episodes and pieces of knowledge that we know, and those we are currently aware of, leads to interesting questions of how these different types of memory are represented in the brain—perhaps some form of neural activity for information in awareness, as opposed to alterations of neural structures for our personal storehouse of facts and experiences.

Most metaphors of memory over the centuries have described the memory system as a storehouse, or palace with many rooms in which memories may be placed. A more up-to-date version of this storehouse metaphor is a large library. In such a system, new books are stored in precise locations according to such specified characteristics as the general topic, date of publication, and author's name. Armed with such knowledge, a borrower can retrieve the sought-for volume successfully at a later time. Even huge libraries like the Library of Congress and the British Library work effectively on this system, but when we consider human memory—when metaphors meet the brain, as it were—the analogy is less persuasive. There is little evidence, for instance, that our millions of specific memories are each stored in just one specific location in the brain; in contrast, most neuroscientists now believe that memories are represented by neural networks distributed widely throughout the cerebral cortex—the brain's outer layers. Additionally, the library metaphor suggests that memories of events and pieces of knowledge are fixed objects, like books on shelves, whereas a stronger case can be made for memories as dynamic *activities* of the brain rather than as static entities, and this is the approach endorsed by most current neuroscientists (chapter 7).

Memory: Not Just "One Thing"

The memory metaphors described above treat memory as being one thing, whereas much recent evidence suggests that there are various types of memory. As an example, in 1917, philosopher Henri-Louis Bergson suggested that the past survives in two forms: as the adjustment of perceptual and motor mechanisms to a particular set of circumstances, and as independent recollections. In the case of a student learning to recite a passage from literature, for instance, Bergson argued that the first reading contributed to the habit necessary to recite the passage by heart in the future and also laid down a memory record of the unique reading of the passage on a specific occasion. The distinction between learned procedures and remembered episodes is a salient part of modern thinking about memory, and we discuss it further below. The general point that memory exists in a variety of forms and functions also makes sense of clinical observations that patients can experience losses in one type of memory while retaining normal function in a second type. We explore such cases in the following sections.

More Than One Type of Memory: Evidence from Amnesic Patients

Various types of brain damage can result in memory impairments and provide evidence on the different forms that memory can take. The findings from patients with brain damage can also shed useful light on memory strengths and weaknesses in healthy individuals, as these normal variations between people (in the case of aging, for example) are often seen in amplified form in patients. A dramatic illustration of the difference between conscious recollection and learned behavior expressed unconsciously was supplied by Swiss neurologist Édouard Claparède. Despite daily interactions with a woman who was an amnesic patient, Claparède had to introduce himself every time he encountered her; on each occasion, she claimed to have never met him before. During one of these encounters, Claparède concealed a pin in his hand and jabbed the pin into the patient's palm while shaking hands. On returning some time later, the patient again failed to recognize him, but when he proposed to shake hands, she refused to do so without being able to say why. She later offered, "Sometimes people conceal pins in their hand!" This anecdote makes it clear that many things we experience, including learned facts and actions, influence our behavior and thoughts though we are not consciously aware of them, or where we encountered them.

A further case of unconscious influences on memory comes from Bruce Whittlesea, a former graduate student at McMaster University. Prior to attending graduate school, Bruce had worked in a hospital housing an amnesic patient whom he frequently met. Bruce saw this as an opportunity to exploit his limited repertoire of jokes. He told the patient one of his jokes and was rewarded by laughter. The following day, he retold his joke, expecting a similar reward. On this occasion, however, the patient did not laugh. When asked whether he had previously heard the joke, the patient said that he had not, but did not laugh because the joke was just "dumb" instead of being funny. The surprise ending required by a successful joke was ruined by its prior telling although the patient did not recollect the prior telling. In subsequent chapters, we provide more instances of how an experienced event can affect behavior in healthy adults, even in the absence of recollecting the original experience.

The Patient Known as H. M.

In 1953, a patient known by the initials H. M. underwent surgery to treat his epileptic seizures. The surgery involved cutting out a neural structure called the hippocampus, one on each side of the brain. The operation produced no change in personality, general intelligence, or perceptual processing, but had the unexpected effect of producing a severe deficit in the ability to record new memories while leaving memory preceding the operation largely intact. H. M. forgot daily events nearly as fast as they occurred as well as the names of people he had just been introduced to. He reported that he often felt as if he had just awakened from a dream.

The results of formal testing were reported in 1957 by William Scoville, the neurosurgeon who had performed the operation, and Brenda Milner, a neuropsychologist.[1] Those results along with later reports showed normal short-term memory, as measured by the ability to hold a small set of numbers in mind, and an inability to form new permanent memories, demonstrated by total forgetting of the numbers after his attention was distracted. Yet H. M. did show preserved memory and learning of other sorts including the ability to perform new perceptual/motor skills. Among those was skill in a task in which he had to trace between two parallel lines outlining the drawing of a star while being restricted to seeing his hand and the drawing only in a set of mirrors, so that his movements to the right were seen in the mirror as movements to the left. His skill improved from day to day, but H. M. had no memory of having done the task before. Thus in line with Bergson's earlier observations, H. M. had a severe deficit in memory for personal experiences, but kept his ability to learn and retain habitual procedures. The further finding that he could retain information in mind, so long as he was not distracted, was taken as evidence that his short-term memory was also intact.

The Role of Prior Knowledge

In the 1930s, the English psychologist Sir Frederic Bartlett proposed that we gradually build up our knowledge of the world from events we experience, and that these experiences are then clustered in organized mental structures he called "schemata." In turn, these schemata (or "schemas") are used to help us understand new experiences and form frameworks in which to remember them. One potential downside of this arrangement is that it is relatively difficult for us to understand and remember information and events that do not fit our current schemata. One of Bartlett's classic demonstrations was to present an unusual North American folktale to an English university student to learn and recall.[2] The student's written recall differed from the original by being shorter and omitting a number of details. This first student's written recall was then given to a second student to learn and recall with the result that more unusual details dropped out of his reproduction, but other details were added, apparently to make the story more coherent and comprehensible to English ears. This procedure was repeated until a series of ten students had learned the previous reproduction and produced their own versions. By the end of the series, the reproductions were much shorter, the supernatural details in the original had been lost, and the whole tale was closer to the experience of English university students in the 1930s. This demonstration thus illustrates the constructive nature of remembering, and effects of beliefs and attitudes on recollection and understanding. Gossip serves as a commonplace example that is similar to Bartlett's findings, with a story progressively changing as it travels across tellings. To return to metaphors for a moment, human memory is *not* like a tape recorder!

Bartlett's experiment shows that prior knowledge can act as a source of errors in remembering. In contrast, work by George Katona, a Gestalt psychologist, found that reliance on prior knowledge can facilitate subsequent memory performance. For example, suppose you were asked to remember the following sequence of digits: 581215192226. The task would be much easier if the numbers were grouped in the following way: 5 8 12 15 19 22 26. Doing so makes it easier to realize that the numbers form a sequence, increasing alternately by 3 and 4. Having noticed this, you need only remember the first and last numbers in the sequence, allowing the remaining numbers to be generated at test. Work done by Gestalt psychologists revealed the importance of principles of organization for both perception and memory.

The Importance of Associations

One striking aspect of human memory is its associative nature; events often remind us of similar ones that happened in the past, and were adjacent in space and time. In this way, a face, song, smell, or scene can evoke memories of related happenings from our mental knowledge base.

These associations typically jump into our conscious awareness automatically, without any effort or intention—our stored memories are apparently highly interconnected. These everyday observations led philosophers and scientists from the seventeenth century on to regard the notion of the "association of ideas" as fundamental to our understanding of the mind. John Locke, David Hume, and John Stuart Mill are some of the well-known names associated with this tradition.

The realization that human memory could be studied experimentally emerged gradually in the latter half of the nineteenth century, especially in the work of the German psychologist Hermann Ebbinghaus. In an attempt to study "pure learning" in the absence of meaning, he used pronounceable three-letter pseudowords such as "DAX," "BOK," and "YAT," strung together in randomly ordered lists; the task was to recall the list of syllables in their presented serial order. For Ebbinghaus, successful learning was attributable to the formation of associative links between adjacent syllables and forgetting was attributable to the loss of such associations. Ebbinghaus was both the experimenter and only participant in his experiments, resulting in the heroic learning of a huge number of lists. To measure learning in a quantitative manner, he recorded the amount of time it took him to learn a list without error, and to measure the retention after an interval of hours or days, he recorded the time to relearn the same list. The

One striking aspect of human memory is its associative nature; events often remind us of similar ones that happened in the past, and were adjacent in space and time.

difference in learning time between the initial learning and this second learning gave a measure of how much had been retained. If the relearning was much faster than the original learning, it may be concluded that the material was well learned and well retained; if the relearning took as long as on the first occasion, we have to say that nothing was retained. Ebbinghaus found that material was lost rapidly in the first few hours after learning, but flattened out and declined more gradually thereafter. This "forgetting curve" is actually well fitted by a common mathematical function; the characteristics of learning and memory obey scientific laws!

What's the difference between "learning" and "memory" in this example? The way these words are typically used by cognitive psychologists and researchers in related fields is that "learning" refers to the gradual accumulation of knowledge about some aspect of the outside world, whereas "memory" refers to the ability to consciously recreate some past experience. So we would say that a laboratory rat *learns* to navigate a maze to get food, but it seems unlikely that it can recollect a particular prior trip or plan a future one consciously before setting out. Similarly, Claparède's patient learned to avoid shaking hands with her doctor, but was unable to remember the details of the original occasion. Do savings in relearning measure memory? They do in the same sense as H. M.'s ability to improve his performance on a mirror-tracing task shows the benefit of practice without being able to report on his earlier experience. The distinction is one between learning and the ability to report on particular prior events— different forms or uses of memory. Following the lead of Ebbinghaus, investigations of *learning* are the ones that were centered on by early investigations of memory in North America.

The Verbal Learning Tradition

This tradition of associative learning was pursued and developed, mostly in the United States, until around 1970. Much of the work was done using laboratory rats as "subjects" (not exactly willing "participants" we suspect!), but parallel studies were carried out on humans using newly formed associative links between unrelated words as material. The participants were asked to learn lists of word pairs such as "donkey-wallet," "harbor-market," "soldier-orange," and so on, with the first word given again at the time of the memory test (e.g., "harbor-?") as the stimulus to evoke the learned response "market." Thus again "learning" was assumed to represent the formation of associative bonds between stimuli and responses, and "forgetting" represented the breakdown and loss of such bonds. Researchers in this tradition were particularly interested in the deleterious effects of interference on forgetting, where "interference" was regarded as the formation of a competing associative link. So in the laboratory experiment with word pairs, when participants were asked to learn a second list with the same first words but a different second word—for example, "donkey-table," "harbor-sunshine," and so forth—the second list "interfered" with the participants' ability to recall the first list, and the first list also interfered with learning the second one.

It was proposed that if the interfering material came after learning, it would "act back" (*retroactive interference*) on the learned material, and if the interference preceded learning, it would "act forward" (*proactive interference*) to make new learning more difficult. A real-life example of proactive interference would be learning French and subsequently trying to learn Spanish; when attempting to come up with the Spanish word for something, the French word may frequently come to mind instead. Similarly, the Spanish learning lessons might interfere retroactively with the ability to retrieve the corresponding French word. Another instance of proactive interference is remembering a friend's new phone number; the old number keeps intruding!

The Cognitive Approach to Understanding Memory

Although the associative "verbal learning" tradition produced many important findings, it was gradually replaced by the cognitive approach that is dominant today. In a praiseworthy effort to put the study of memory and learning on a rigorous scientific footing, researchers in the associative tradition had concentrated exclusively on the observable and measurable characteristics of materials, learning conditions, and participants' responses. The emphasis was entirely on *behavior*. This had the consequence that ideas of mind and consciousness were ruled out of consideration, along with emotional reactions, subjective strategies, and personally meaningful images. Starting in the 1950s, the cognitive approach stressed the importance of these internal feelings and mental states, and also the significance of *organization* for memory—the idea that people learn and remember facts and events not by forming associative links between items but rather by clustering them into meaningful groups—much as Bartlett had suggested twenty years previously. Our existing knowledge provides a meaningful structure that acts as a sort of "mental scaffolding" to aid both learning and retrieval.

The central role of organization in memory was underscored by many researchers working in the cognitive tradition. Experiments showed that when people are given a long list of randomly arranged nouns to learn on several successive learning trials, they gradually impose a personally meaningful "subjective organization" on their recalled responses. So, for example, if the words "horse," "barn," "tractor," and "chicken" are scattered throughout a list, learners might form the category "things found on a farm" in their mind, and group the terms together when they recalled the list later. Such mental groupings can be quite unique to individual learners—for instance, "objects encountered on my way to the office today" or "things in my living room." At the time of the recall test, the participants first attempt to retrieve the categories they have formed and then the words nested under each category heading. Whereas workers in the verbal learning tradition had ignored memory for *events*, focusing largely on optimal conditions for effective learning, memory for particular episodes became the focus for much research in the cognitive tradition. In this vein, research on eyewitness testimony is described in chapter 5.

Short- and Long-Term Memory

Cognitive psychologists have proposed a number of ways to dissect human memory into different forms reflecting different functions and their failures. One obvious distinction is between information currently held in mind and information that we know, but that must be retrieved from a large store of personal experiences and general knowledge. The first type is commonly referred to as short-term memory, and the second type as long-term memory, although these terms have been used rather differently by different professionals and by the general public. The way we will use them is to define short-term memory simply as the function that enables us to hold a small amount of information in conscious awareness at a given time; all other information is in long-term memory by this view. We will go over the evidence for this distinction in further chapters, but just to emphasize the point, short-term memory is the system we use to hold an idea, telephone number, name, image, or musical phrase in mind in order to respond appropriately to a question or carry out an appropriate action.

The features that differentiate short-term from long-term memory are its limited capacity, vulnerability to forgetting, and the way in which information is stored. As an example of this last aspect, verbal material is typically held in short-term memory in the form of inner speech, whereas the *meaning* of words, phrases, sentences, and stories is more salient in long-term memory.[3] We know from our experience that short-term memory is limited; a seven-digit local telephone number is about most people's limit. On the other hand, no one has ever demonstrated a capacity limitation on material stored in our long-term memory. At first that statement seems hard to believe. The brain is just a physical organ after all; surely there must be a limit to what it can retain.

One way to think of it is in terms of the possible *combinations* among neural elements. The English language

Short-term memory is the system we use to hold an idea, telephone number, name, image, or musical phrase in mind in order to respond appropriately to a question or carry out an appropriate action.

involves only twenty-six letters, but we use them in many different combinations to form somewhere around fifty thousand different words—or a lot more if we include acronyms and imported non-English-language words. The point is that with an estimated eighty-six *billion* neurons in the brain, the number of combinations to represent words, images, ideas, and records of our experiences is essentially unlimited. Of course we forget the names of people, places, and objects all the time as well as forgetting personal experiences and intentions, yet these failures of memory are most likely failures to *access* information that is still present in our long-term memory in some form. We know this from the everyday experience of a forgotten name suddenly "popping back" into our conscious awareness minutes or even days later. Another piece of evidence on the same point is that after failing to *recall* half the items on a long list, the person will probably *recognize* the missing items when presented with a further list containing the forgotten words, or objects scattered among new words or objects. The information was still there, but it required a better retrieval cue to be recollected.

So while forgetting from short-term memory entails displacing the information to attend to something different, forgetting from long-term memory is usually a case of an inability to retrieve the wanted information, although it is still there in some sense. As we discuss further in chapter 2, the key to successful retrieval from long-term memory is to provide the person with an appropriate reminder or retrieval cue. Usually such cues involve the context in which the information was learned or the intention was laid down. Think of forgetting what you came upstairs to fetch; you often have to go back downstairs in order to be reminded. We once heard a radio comedian observing that while sitting in his armchair in the living room, he frequently intended to fetch some article that he needed upstairs, but after arriving upstairs totally forgot his intention, only to remember the article when he sat down again in his living room chair. On the basis of this evidence, he concluded that "memory is in the ass!" Although physiologically implausible, the comedian's theory actually has some merit! The total context of his living room, including his bodily sensations, form the context—the environmental support—in which the intention was embedded, and this context needs to be reinstated, in part at least, for the intention to be remembered.

To summarize this distinction between short- and long-term memory, the former term refers to the limited amount of information in our present conscious awareness; the information is lost when we must attend to something else. Long-term memory is "memory proper"—the enormous, semipermanent system in which facts, experiences, and accrued knowledge are all retained. Knowledge is typically well organized, and the difficulties of memory retrieval are reduced when the learned material is compatible with these organized forms. As we explore later, the general context of an experienced event (involving place, present company, time of day, and mood) is typically laid down or "encoded" as part of the experience, and such features later serve as effective retrieval cues to aid in the reconstruction of the original event.

Memory Systems

The Canadian psychologist Endel Tulving and his associates have proposed that human memory is composed of five semi-independent systems, each with its own characteristics.[4] The two major systems are, first, one to retain events that have happened to us personally—*episodic memory*—and second, one to hold abstract knowledge—*semantic memory*. So episodic memory is the system associated with "memory" as we typically use the term—remembering what we ate for breakfast yesterday morning or the details of that wonderful vacation we spent in Paris five years ago; "autobiographical memory" is another term that researchers use to describe this ability. Tulving has also coined the phrase "mental time travel" to capture the idea that when remembering some past event, we can reexperience its sights, sounds, and moods to some degree (sometimes erroneously, as we will discuss in later chapters!).

In contrast, semantic memory holds facts, names, locations, and general knowledge of the world. One way of describing the differences between episodic and semantic memory is that episodic memories involve contextual features such as time and place, whereas information in semantic memory is abstract and context free. As an example, we can draw on semantic memory to describe a typical bicycle as having wheels, pedals, a chain, and so on, but we must access episodic memory to relive the occasion when we were chased by a dog while on a cycling trip in the countryside last week. The evidence for the difference between these two types of memory comes partly from observations of patients who show impairments of one type of memory while displaying normal behavior on the other type. An important patient, studied by Tulving and his associates, was a young man who had sustained severe brain injuries as the result of a motorcycle accident. The patient, known by his initials K. C., had a dense amnesia for life events, but could answer factual questions about a variety of topics and everyday procedures.[5] For example, he could tell you in some detail how to change a tire on an automobile, but when asked to recall some occasion when *he* had changed a tire, he came up blank.

Interestingly, when K. C. was asked what he planned to do tomorrow or where he might spend a vacation next month, his mind was equally blank; he described his attempts to do so as like looking into a large empty room with no furnishings. So it seems that the processes involved in retrieving memories of past events—episodic memory—are also used to envisage upcoming events and in planning for the future. In fact, some theorists have speculated that the ability to free a person or primitive animal from experiencing only what is happening at present—the "here and now"—may have evolved *primarily* to enable animals to plan their next moves and avoid some danger that threatened them on a previous similar occasion.

Both episodic and semantic memory may be regarded as forms of long-term memory in the sense that both types of information have to be retrieved in order to be in our conscious awareness. Information that we are holding in mind constitutes a third system in Tulving's scheme; this type of short-term retention is referred to as *working memory*, and we describe it more fully in chapter 2. His final two proposed systems are *procedural memory*, the ability to retain learned procedures like riding a bicycle, writing longhand, or playing golf or a musical instrument, and the *perceptual representation system*, referring to the set of processes through which sensory systems are tuned to make finer discriminations among incoming sense data through sensory interactions with the environment as well as through practice in performing specific sensory tasks. Examples include developing abilities to distinguish among several odors, musical pitches, or different shades of colors.

Automatic and Controlled Processes

In simple animals, many aspects of behavior are controlled by genetically mediated innate response tendencies commonly known as instincts. Such behaviors are common to all members of a species, unlearned, and typically triggered by specific stimuli. The problem with such automatic patterns of behavior is their lack of flexibility in unusual circumstances so that as animals evolved, control was gradually shifted from external triggers to internal consciously controlled mechanisms of choice. At the human level, many habitual behavior patterns are still run off automatically, although they are usually learned rather than innate. Habitual procedures require less attention and fewer "processing resources" than do consciously controlled procedures, so we tend to revert to this "automatic pilot" mode when we are tired or our attention is distracted. The balance between habit and control is seen in many human memory situations, and some studies illustrating this balance are examined in chapter 4.

Control is mediated by processes located in the frontal lobes of the brain. In general terms, the frontal lobes regulate brain functioning by selecting, initiating, and maintaining control over other brain processes—especially those concerned with such higher cognitive functions as thinking, planning, and problem-solving. These management processes are referred to as *executive functions* and come into play whenever nonroutine situations require novel solutions. This type of cognitive control is involved in the strategies that a person may employ during learning and memorizing—by consciously organizing the material, for example, or relating events to their contexts in rich and meaningful ways. Executive processing is also involved during retrieval processing when individuals make effortful attempts to elaborate and integrate the initial fragments of memories they are trying to recover. Finally, executive functions are again involved when information in mind (in working memory; see chapter 2) must be manipulated and transformed.

These executive control functions are illustrated by the difficulties experienced by patients with frontal lobe damage. Such patients have problems with thinking, planning, and deciding, and also show impairments of memory encoding and retrieval. In the case of retrieval, the problems are greatest in "free recall," where the person attempts to recollect some event in the absence of hints or reminders. Such unaided recollection involves reconstructive processing, which in turn requires executive control. On the other hand, recognition memory is less affected by frontal lobe damage, as in this case the re-presented information provides fuller cues to aid in the retrieval of the original event. The frontal lobes are among the first parts of the brain to deteriorate in the course of normal aging, and so we would expect some forms of memory to decline but other forms to remain unchanged in older adults. We look at work on this topic in chapter 6.

FROM STORES TO PROCESSES

When computers emerged in the 1950s, they provided an obvious analogy to brain function. Both computers and brains receive inputs from the external environment, process that information in terms of the knowledge they already possess, store it both temporarily and permanently, make decisions, and respond in light of the processing just carried out. Cognitive psychologists were struck by these parallels, and so constructed models (descriptive frameworks in verbal, pictorial, or mathematical terms proposing components of human memory functions along with the relations among them) that captured what was known about how humans process and store information. These aspects included the points that there is a strict limit on what humans can attend to and therefore process at any one time; it takes some time to comprehend the information in terms of what we know, and further time to decide on an appropriate course of action. It seems we can accumulate huge numbers of facts, learned procedures, and memories of events we have experienced, almost all of which we are unaware of at a given moment, but can retrieve via hints and reminders provided by ourselves, others, or the environment itself. In addition to this extremely large long-term store, we can remember a much smaller amount of information (e.g., names, numbers, and intentions) by holding the information temporarily in mind, in conscious awareness.

Models of human memory proposed in the 1950s, 1960s, and 1970s thus focused on two major memory stores: a short-term store with an extremely limited capacity, and a long-term store with an infinite capacity as far as anyone knew. The material in the short-term store corresponded to the information in conscious awareness, and therefore both preceded and followed the registration of information in the long-term store. That is, when learning new information, we are first aware of the material to be learned by seeing or hearing it; the material is then transferred into our more permanent long-term store. But when we recollect some previous experience or wish to review some previously stored fact, we retrieve it from the long-term store so that it is again part of our conscious experience.

As an example of such a scheme, US psychologists Richard Atkinson and Richard Shiffrin proposed a model of how incoming information is first registered and processed by the sensory systems, and then transferred by the processes of attention to a limited-capacity short-term store.[1] Here the information may be retained in conscious awareness by *rehearsing* it continuously. The material in short-term store can be manipulated and transformed (think of mental arithmetic or solving an anagram mentally), and also copied into a long-term memory store of much larger capacity for semipermanent retention. Given appropriate retrieval cues, the material can be recovered later and reentered consciously into the short-term store. This is the type of short-term memory that we described as being intact in the amnesic patient H. M. in chapter 1. The Atkinson and Shiffrin model of memory thus focuses on certain aspects of the system, prompting questions that may be answered by experiments to test the ideas. Such questions include, What is the mechanism of transfer from one store to the next? How is the material represented (or "encoded") in different stores? What is the capacity of each store? And how is information lost or forgotten? The model is shown below.

Sensory Memories

The first memory box in the Atkinson and Shiffrin model (the sensory stores) reflects the idea that information

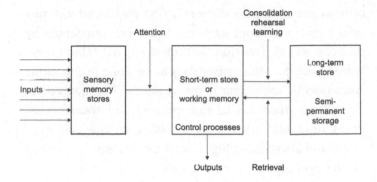

Figure 1 Information enters the memory system via the senses, where it is held briefly. If attended to, the information is then entered into the short-term store, and while there it may be copied into the long-term store for permanent storage. At some later time, the information can be reactivated and reentered into the short-term store to play a role in thinking and decision-making. *Source*: R. C. Atkinson and R. M. Shiffrin, "The Control of Short-Term Memory," *Scientific American* 225 (1971): 82–90.

from the environment—stimulation to the eyes, ears, skin, and tongue—is held briefly in a raw sensory form before being processed more deeply for identification and significance. In the case of hearing, we can get some idea of what's meant by auditory sensory memory in situations where your concentration is strongly focused on *visual* information, such as reading a book perhaps or watching an engrossing TV series. If a significant auditory event occurs at this moment—a clock strikes or a companion asks you a question—the sensation is that you can turn your attention to the mental record of the sound and count the number of clock chimes, or understand the question that was asked. This type of rapidly fading trace is known as *echoic memory* for obvious reasons; experiments have shown that it lasts around two seconds. If we believe that the information is important, it is necessary to "rescue" it from this sensory record by the processes of attention and so transfer it to the short-term store, where we are aware of it and where it may be maintained indefinitely as long as it is continually refreshed by repetition.

Similar fleeting memory systems are associated with other senses. In the case of vision, the corresponding system is referred to as *iconic memory*; experiments have shown that the visual system initially registers a huge amount of information from the environment, but the resulting sensory record fades extremely rapidly—in about one-fifth of a second. Why should iconic memory fade so much faster than auditory sensory memory? One reason is that visual information in the environment is relatively stable; if we don't perceive an object fully at first glance, we can usually look again. In comparison, auditory events are transient (e.g., the sound of a twig snapping in the forest); we therefore need more time to register and analyze them. Also, auditory events often unfold over time (e.g., words in a spoken sentence or a melody); in these instances, the earlier parts of the sequence must be held briefly before being integrated with the later parts to comprehend the meaning. Corresponding sensory memories are likely present for touch, smell, and taste too, although they have been studied less than those for vision and hearing.

Short-Term Memory

The concept of short-term memory is familiar to most people, and is typically used in the sense of memory for recent events. Older people frequently complain that whereas their memory for events experienced earlier in their lives is excellent, their short-term memory is dreadful! This usually means that they now tend to lose their car keys, forget what they came to the fridge for, forget that they already shared some gossip with a friend, and even fail to keep appointments. As we will discuss later, these annoying memory failures are due to less effective registration (encoding) of new information, less effective retrieval processes, and impairments in prospective memory—a failure to recollect as well as act on plans and intentions that were made minutes or even seconds previously. Yet these commonplace difficulties when dealing with recent events reflect problems of *long-term memory* in the sense used by Atkinson and Shiffrin; the short-term store in their model has rather different characteristics. The label *short-term store* essentially characterizes the small amount of information we can hold in mind at any one time, so it is the kind of memory we use to retain an unfamiliar telephone number or mail code before dialing it or writing it down.

One major difference between the short- and long-term stores in the Atkinson and Shiffrin model is the extremely limited capacity of the former. Despite its small capacity, however, it is a crucially important component of cognition, as we describe under its later designation of *working memory*. The second striking difference between the stores is how rapidly material is lost from the short-term store. Experiments showed that when young adults were given three meaningless syllables (e.g., *MOR*, *TAK*, and *LIN*) to remember briefly, performance was high if recall followed the presentation immediately, but when experimental subjects were given a second demanding task to perform before recall (e.g., mentally subtracting threes from a given number, as in ninety-eight, ninety-five, ninety-two, eighty-nine, etc.), memory for the syllables dropped from near 100 percent for immediate recall to around 10 percent after only twenty seconds of continuous subtraction. Again this finding contrasts with the long-term store, which seems to retain well-registered information for lengthy periods and is relatively resistant to distraction.

A third difference between the stores is the manner in which verbal information is represented in terms of speech sounds in the short-term store, but predominantly in terms of meaning in the long-term store. The evidence for this difference came partly from the types of error that the experimental participants made when recalling items from the two stores. When errors were made in the immediate recall of a string of letters, such as $A\ J\ B\ R\ S\ Y\ K$, the errors often sounded like the correct letter; that is, P instead of B, or F instead of S, in the present example. This result was in contrast to items retrieved from the long-term store; when the participants made an error when recalling a list of words presented fifteen minutes ago, the error was most likely to have a meaningful connection to a word on the list. They might recall "tiger" instead of "lion," say, or "cousin" instead of "nephew."

The conclusion was that verbal information in the short-term store is encoded as speech sounds, whereas the same material is held in the long-term store in terms of meaning. But just to complicate the picture, it turned out that meaning can also play a part in how well information is held in the short-term store. If groups of letters make up well-known acronyms, such as SOS, DNA, NBC, and CIA, experimental subjects can recall almost as many groups (or "chunks" as they are referred to[2]) as they can recall individual letters. It is also the case that most adults can hold and recall a set of fifteen to twenty words if they form a highly meaningful sentence; try repeating "*The small bearded man limped slowly across the sunlit courtyard and picked up an abandoned pack of sandwiches.*" The contrast between this result—eighteen words—and the limit of around only five words if they are unconnected nouns, like "cloud," "happiness," "table," "arithmetic," and "lizard," is dramatic, and clearly means that the short-term store is not simply an "input buffer" for holding incoming information before it can be processed for meaning and implication. Rather, it seems that the short-term store can use the knowledge we possess and retain the results for whatever action is required. So the material in the short-term store—held in conscious awareness—may come either from information we have just perceived from the outside world, or information retrieved from our permanent store of facts, knowledge, and events.

On the question of how information is transferred from the short-term to the long-term store, the original suggestion was that this occurred automatically as a function of the length of stay in the short-term store. Later research, though, showed that mere residence in the short-term store and even repetitions of items held in short-term store are not effective methods for building good representations in long-term memory. On the other hand, processing the information semantically *is* effective, in line with the levels-of-processing principles discussed later. Simple repetition can be effective for later recognition memory, however, and in chapter 3 we look at how repetition can be quite effective when the repetitions are spread far apart.

In the Atkinson and Shiffrin model sketched in figure 1, you will see that the box labeled "short-term store" is labeled "working memory" too, and contains the mental activities of rehearsal, elaboration, and manipulation. So the theorists' suggestion was that the short-term store is not simply a passive store to hold items of information but also a mental workspace in which complex cognitive operations are carried out. This switch in perceived function gradually caused researchers to drop the term "short-term store" and refer to this more dynamic construct as *working memory*. We describe some research on working memory in the next section.

Working Memory

The concept of working memory was introduced in 1974 by British psychologists Alan Baddeley and Graham Hitch.[3] Their model involved stores to hold recently perceived visual and acoustic (or phonological) information, and an *articulatory loop* to capture the idea that individuals can recirculate short phrases, lists of words, letters, or digits by rehearsing them mentally. Processing operations in these various peripheral stores are managed by the *central executive*; basically a system of cognitive control mediated by processes in the brain's frontal lobes. Working memory is a more flexible and dynamic way to talk about the short-term retention of small amounts of material; it stresses the cognitive *work* being carried out on that material as opposed to the relatively passive notion of memory stores.

This concept has greatly influenced models of memory during the past fifty years, and has also evolved in theoretical schemes put forward by the original researchers and others. One relatively recent addition to the working memory model is the *episodic buffer*, which holds and integrates various types of incoming information (e.g., auditory, pictorial, temporal, and contextual), and binds them together to enable the conscious apprehension of a single event.[4] Among the many practical applications that the model has engendered is the involvement of the phonological store in vocabulary acquisition and second-language learning. Many recent experiments have explored the characteristics of *visual* working memory, such as by displaying a set of differently colored shapes briefly and then showing a second display after an interval of anything from half a second to several seconds. The second display may be identical to the first, or contain an altered shape or color. Such experiments have demonstrated how long different kinds of visual information can persist in working memory.

Other researchers have accepted the basic idea of a working memory system, but have interpreted the findings differently. The version we will discuss here suggests that working memory basically involves attention allocated either to new incoming perceptual information or representations of words, facts, images, concepts, and memories that already exist in our knowledge systems.[5] By this notion, information "in working memory" is simply information that we are attending to—an idea that converges with our experience that attention and conscious awareness are closely intertwined. The defining characteristics of short-term memory—limited capacity, rapid forgetting, and the "acoustic coding" of speech information—fit well with this new perspective. Limited capacity is now seen as a limit on *attention* as opposed to a limited space in which to hold items; if we are paying close attention to a newspaper article, we do not immediately comprehend a remark spoken by a companion. In the same vein, when individuals wearing headphones are given a task demanding close attention to information presented to one ear, they can report little about information played to the other ear. If the person's own name is presented to the unattended ear, however, it *is* perceived by some participants, showing that highly meaningful material can break the attention barrier. Interestingly, individuals with strong working memory abilities do *not* hear their name. The difference is thus not in overall working memory capacity but instead in cognitive control; high working memory abilities enable a person to resist interference, making them less distractible and so less aware of even important signals such as their own name.

Rapid forgetting from working memory is accounted for in terms of switching attention to some different source of information; the items that we were attending to (e.g., words in the newspaper article) drop from our awareness and are replaced by our companion's remarks. The type of code used in working memory is simply the particular dimension attended to (e.g., visual words or auditory remarks). As in previous models, speech information, including numbers, letters, words, and sentences, appear to be retained by attending to and rehearsing the appropriate speech sounds—an output code rather than a code representing recent inputs or stored meanings. A further benefit of this way of looking at working memory is that the type of code used is flexible by definition—it just depends on what we are attending to. As well as retaining verbal information, we can retain images of visual patterns and musical phrases, and even information about touch, taste, and smell to some extent.

The idea of short-term memory has therefore progressed from a mental device to hold a brief series of words or numbers, to a concept dealing with the central purposes of intelligence and cognition. Working memory is the mental workspace in which ideas can be integrated with other ideas and new incoming information. To perform these crucial cognitive activities, information must be held in mind while it is processed, but in a sense this memory aspect is secondary to the other cognitive operations being carried out. Working memory may be regarded as the interface between the cognitive system and the external environment, and as such is a central aspect of cognitive functioning.

A Processing Approach

The idea that memories exist as "things in the head," as structures located in stores of limited capacity, has intuitive appeal. It *feels* as though we have to search through a complex space to find a name or fact, or the details of a previous event. Much current research, however, suggests that memories are represented by networks of neurons distributed broadly throughout extensive regions of the brain. This point, along with inconsistencies in the memory stores story, led researchers to think of memory in a different way: as a set of activities in the mind and brain rather than as a collection of structural traces— that is, "remembering" as an active verb versus "memory" as a static noun.

Attention and memory were also brought together in the "levels of processing" framework proposed by Canadian psychologists Fergus Craik and Robert Lockhart in 1972.[6] Previous work by British psychologist Anne Treisman had suggested that incoming perceptual information is subjected to various levels of analysis running from peripheral sensory analyses of brightness, colors, sounds, and smells, through the identification of visual forms, voice qualities, and sound frequencies, to the perception of objects, words, and melodies, and eventually to the deeper meaning and significance of what we are perceiving. The ability to perceive these final, deeper levels is tied up with attention in the sense that in general, more attention is required for deeper levels of analysis. Early "shallow" analyses of sensory information are automatic—we are aware of the voices of other people in the room without paying attention to them—but attention is required to understand the meaning of the information we are hearing or seeing. So attention is a necessary component of Treisman's "levels of analysis" view of perception.[7]

The essence of Craik and Lockhart's notion was that memory could be added to this mix of attention and perception. Their proposal was that information is registered and retained in memory to the degree that it is processed deeply and meaningfully. So the term "depth of processing" refers to the extent that incoming information has been analyzed, with deep, meaningful processing associated with long-lasting memories. In addition to this depth factor, the authors later suggested that good memory also depends on the degree of richness and elaboration achieved when encoding the event. In the case of memory for words, one experimental participant might register only that "jackal" is an animal name, whereas another learner might reflect that they had recently seen a nature film on TV in which jackals were seen devouring the carcass of a deer while making yipping and barking noises. The greater amounts of elaboration—pictures, sounds, and associations—constructed mentally by the second participant will make the word better remembered in a later test.

Another important factor is how *distinctively* an event is encoded in the memory system.[8] Just as a distinctive feature in a visual scene—a white golf ball on a bright green fairway, for example—can be easily located and perceived, so a distinctively encoded event will be well located and retrieved. The factors of elaboration and distinctiveness are often related, especially if the further elaborations act to make the object different from other objects of the same type. In turn, the elaborating details that a person draws on will come from the person's relevant knowledge of that object or event—from the knowledge schemata proposed by Bartlett (chapter 1). When the memory is sought, elaborate encodings provide more opportunities for its successful retrieval. Also, to the extent that an encoded event is well integrated into a schematic knowledge base, the knowledge can be used as a "scaffolding" to organize effective retrieval processes. In summary, according to the levels-of-processing framework, good memory is a function of the depth of processing, elaboration, and distinctiveness of the encoded representation at encoding. Retrieval is a question of reconstructing the same set

The term "depth of processing" refers to the extent that incoming information has been analyzed, with deep, meaningful processing associated with long-lasting memories.

of mental processes that were involved at encoding, and these retrieval operations are facilitated to the degree that the encoded representation is congruous with the person's schematic knowledge base.

These ideas were explored in a series of experiments reported by Craik and Tulving in 1975.[9] Their basic idea was to induce the experimental participants to pay attention to aspects of words at various depths of analysis ranging from their shallow surface characteristics to deeper meaningful aspects. After an initial registration or "encoding" phase, the participants were given a memory test for the words, with the prediction that the deeply encoded words would be remembered better than those encoded in a shallow manner. In one typical experiment, the participants were shown a series of 60 concrete nouns, one at a time, and given a question to answer about each noun before it was flashed briefly on a computer screen. The questions concerned the typescript (e.g., "Is the word printed in capital letters?"), sound (e.g., "Does the word rhyme with 'X'?"), or its meaning ("Would the word fit the following sentence frame?"). These three levels of processing—case, rhyme, and sentence—were intended to induce the participants to process different words to increasingly deep levels of analysis. Half the questions led to a "yes" answer and half to a "no" answer (e.g., "Does the word rhyme with 'bread'?—'CLOUD.'") The participants were informed that the experiment was to see how rapidly they could answer different types of question about words; they were not informed of a later memory test. Following the 60 questions and answers (each with a different noun), the participants were unexpectedly given a typed list of 180 words—the 60 they had answered questions about mixed randomly with 120 similar words. The task was to recognize the 60 original words.

The results are depicted in figure 2. The left panel illustrates the speed of responding ("response latency") to each type of question. The scale is in milliseconds, where 500 milliseconds equals 0.5 seconds. The graph shows that responses to sentence questions took longer than those to rhyme or case questions. The right panel portrays the different levels of recognition memory for each type of question. The scale gives the proportion of correct responses for each experimental condition, so for words presented with the question "Was the word printed in uppercase letters?" the proportion recognized was 0.15 or 15 percent. For words that did fit the sentence frame presented in a "sentence" question, the proportion recognized was 0.82 or 82 percent. The recognition memory data thus demonstrate that words in the "sentence-yes" condition were recognized five times more than words in the "case-yes" condition. Simply answering a question about the word's meaning, compared to a question about its typescript,

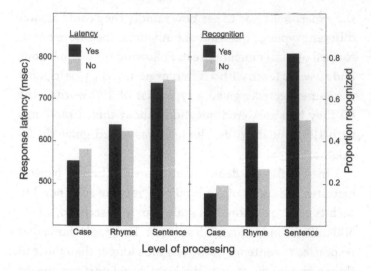

Figure 2 The left panel shows the response times to the different kinds of questions about nouns; for example, sentence questions were responded to in about 740 milliseconds (0.74 seconds). The right panel shows the later recognition performance for the nouns; for instance, rhyme-yes nouns were recognized on 48 percent of the occasions, and sentence-yes nouns on 82 percent of the occasions. Reproduced from F. I. M. Craik and E. Tulving, "Depth of Processing and the Retention of Words in Episodic Memory," *Journal of Experimental Psychology: General* 104 (1975): 268–294, with permission from the American Psychological Association.

increases its memorability by a factor of five. Deep processing is the key to good memory!

One unexpected result was that "yes" responses were better recognized later than were "no" responses. This difference was attributed to the more elaborate or richer encoding achieved when the test word could be integrated with the question. So for the sentence "The boy bought a ___ at the store," the following word "COMIC" fits the sentence (and therefore a "yes" response should be given), whereas the following word "CHURCH" does not. In the next section, we will discuss other factors underlying the successful encoding of events into this semipermanent system, and also the other side of the coin: the factors underlying successful retrieval.

Encoding and Retrieval Factors in Long-Term Memory

We use the word "encoding" to describe the registration of information into memory and how it is represented. So, for example, we talked earlier about words in short-term or working memory being encoded in terms of inner speech, but in long-term memory in terms of meaning. Obviously we remember many other types of material apart from words. We remember pictures, scenes, faces, voices, smells, tastes, and many other aspects of our world, but the common feature seems to be that literal copies of perceived information are integrated in the brain with other instances of the same objects, faces, or events to form higher-order representations of the many individual occurrences. In turn, this accumulated knowledge of the world serves to interpret further instances of the same perceptual event. Hence we interpret the world in terms of what we know, and new information is laid down in memory colored by our individual experiences and beliefs. We give some concrete examples later in the book.

In chapter 1, we made the point that while the notions of limited capacity and rapid forgetting are good descriptions of working memory, the terms are much less appropriate for long-term memory. As we mentioned previously, no one has ever demonstrated a limit on the amount of material that can be held in long-term memory. The system in fact seems less like a box that can hold only so many items and more like a complex system of scaffolding—the more knowledge we have about a topic, the easier it is to add more knowledge and remember events related to the topic. We discuss this central role of expertise in memory in a later chapter.

With regard to forgetting in long-term memory, a number of memory theorists have suggested that once a fact or event is registered in memory, it is there forever—and there is *no* forgetting! At first this sounds absurd; clearly we cannot remember trivial details of daily events from a year ago. There is an important distinction, however, between information that is still "in memory" in some sense and the ability to retrieve it for conscious awareness. Information from previously experienced events clearly plays a part in shaping our current understandings and beliefs about the world although we are rarely able to retrieve these events as episodic memories.

A number of memory
theorists have suggested
that once a fact or event
is registered in memory,
it is there forever—and
there is *no* forgetting!

As an analogy, consider that some mythical person who (for reasons best known to himself!) is creating a beach by depositing trillions of pebbles and grains of sand, one at a time over a lengthy period. Each pebble or grain (representing an individual event) can be perceived and described as long as it is still in the person's hand (i.e., in working memory), but once deposited, it is lost among the many millions of similar grains. Nonetheless, the accumulating grains do create a distinctive beachscape that symbolizes the person's accumulated experiences. Also, if a specific pebble is large and different enough, it can still be identified and picked up again by our strolling beach maker. That is, distinctively encoded events are more likely to be successfully retrieved.

To be less fanciful, it is certainly the case that distinctive events can be retrieved more readily from long-term memory and reexperienced as episodic memories. A distinctive event could be something unexpected and shocking such as the tragic death of a public figure, or it could be a memorable family occasion or incident in which you receive public praise for some action you have taken. In general terms, the key to understanding how well events are remembered involves, first, processing the event in a meaningful, distinctive manner, and second, providing some aspect of the encoded event at the time of retrieval to serve as an effective reminder or "retrieval cue." When we experience any complex event, many aspects or features of the occasion make up the context in which the event was experienced. Say we meet a friend unexpectedly on a street corner. What we encode about the event involves not only the friend's appearance and the physical surroundings of the meeting but also our feelings toward the person, weather, time of day, our own background mood, and many other small details. These experienced features are all encoded in memory as part of the event; how distinctive the encoding is will depend largely on how similar these features are to many millions of other encoded events. If we encountered the hometown friend unexpectedly when entering a hotel elevator during a trip to Singapore, this would make the event more distinctive—and more memorable—than meeting the person on the street where we both live.

Successful retrieval consists essentially of using some of the event's encoded features as cues to access the stored event and hopefully reconstitute many of the remaining features. Two points in this connection are, for one, that successful retrieval depends on how many of the encoded features are available at the time of retrieval, and second, how specific (or "diagnostic") the retrieval information is to the encoded event. The first point may be summarized as "reinstating the context" in which the event occurred (see also chapter 3); so if on a second visit to Singapore three years later you are about to enter the same hotel elevator, this may well remind you of your previous meeting.

"Revisiting the scene of the crime" to help witnesses recollect details of experienced events is a further example.

With regard to the second point concerning the specificity of the retrieval cue, it makes sense that if a potential reminder is relevant to many other encoded events, it will not be effective; clearly the cue "friend" will not be as effective as the cue "friend, elevator, Singapore" to evoke the wanted memory. So for a retrieval cue to be successful, the information in the cue must be part of the encoded record and should be as specific as possible to the memory you are trying to retrieve.

In terms of general principles then, we can say that encoded information about an event involves many contextual features as well as the central items of interest, and that these contextual features will later serve as effective reminders for the event to the extent that they are specific to the sought-for information. In the same way, a memory will not be retrieved if the cue is either not part of the encoded record or is relevant to many other events in memory. As such, much forgetting in long-term memory—perhaps *all* forgetting in long-term memory—is "cue dependent" in the sense that the information is still there, but the cues and reminders are not sufficient to evoke the memory.[10] This way of looking at retrieval as a set of interacting processes highlights another major point about remembering: retrieval is *reconstructive*. That is, just as initial perception involves information from the environment shaped and interpreted by our existing knowledge, so successful retrieval is a matter of integrating the cue information with relevant records in memory, again using our knowledge to shape and reconstruct the original event as fully as possible. The last and possibly most important general point is that retrieval is therefore very largely a question of repeating the mental operations that took place when the event was initially perceived and understood. Retrieval processes recapitulate encoding processes. As we will discuss in chapter 7, there is now good evidence from brain-scanning experiments to corroborate this principle.

COMMON PROBLEMS AND GOALS
More on Encoding and Retrieval

Effects of Expectancy, Needs, and Choice on Memory Performance

The constructs of attention, perception, thinking, and memory were all traditionally regarded as independent faculties of mind, but by the 1960s it was becoming clearer that they are actually interdependent components of one larger cognitive system. What we attend to, perceive, and think about is recorded in memory too. It was also realized that computer models of memory do not take into account the more subtle aspects of human thought, such as cultural background, individual interests, and needs that play a part in how people perceive and remember aspects of the world. Jerome Bruner argued for the importance of such factors.[1] In one classic experiment, he asked ten-year-old children to adjust the size of a light circle shown on a screen to match the sizes of coins of different values ranging from one to fifty cents. The children held the actual coins in one hand while adjusting the size of the circle. All the children overestimated the size of the coins relative to cardboard discs matching the coins in size, but this overestimation effect was greatly amplified in children of poor parents relative to children of rich parents. So need affected the remembered size of coins in this study, even when the coins were perceptually present. Although there have been some failures to replicate this result, the finding does serve to show the close correspondence between the effects of needs on perception and those on memory.

In another surprising demonstration, researchers asked the experimental participants to recall or identify details of lettering and orientation on the US one-cent coin.[2] Most of the participants were unable to do so despite the fact that it was an object they had seen and used on hundreds, if not thousands, of occasions. Why was memory so poor for the details of such a familiar object? In the previous chapter, we stressed the importance of meaningfulness for memory, yet the coin was certainly a meaningful object to the people in the experiment. The researchers suggested that the details, while meaningful, are not *functionally useful* in discriminating the penny from other coins. In Canada, banknotes of different value are colored differently, and each valued note has a distinctive picture on one side of the banknote. Memory for pictures is typically excellent, as we discuss later, yet individuals in an informal Canadian study were unable to recollect the pictures on various notes. In contrast, people were extremely good at remembering the notes' colors. Again, the critical feature seems to be functional usefulness; color is used routinely to select appropriate notes, whereas the pictures are irrelevant to that selection.

Sometimes memory is so poor as to result in blindness to change. A dramatic example of change blindness comes from a study in which a pedestrian on a university sidewalk was stopped by a person (actually an experimenter) dressed as a construction worker carrying a map and asking for directions.[3] While they were talking, two other "workers" walked between them carrying a large door. While hidden behind the door, the original questioner was replaced by a different experimenter also dressed as a construction worker. The two were dressed differently, varied in height by approximately five centimeters, and their voices were clearly distinguishable. After appearing from behind the door, the substitute construction worker resumed asking for directions. The goal of the study was to determine whether the person asked for directions would notice the change in the questioner. After giving directions, the pedestrian was asked, "Did you notice anything at all unusual after that door passed a minute ago?" Surprisingly, only about half the pedestrians noticed that the experimenter had changed, and all of those who noticed change were students who were roughly the same age as the experimenters. Those who failed to notice the change were substantially older than the experimenters. This difference likely reflects greater attention being given to one's own peer group. The surprising result of the experiment shows how the strong expectation of continuity can influence perception and memory.

Measuring Memory

How is memory measured in the laboratory? Researchers use different methods depending on the materials and the research question under investigation. Perhaps the most straightforward method to measure episodic memory—memory for personal events—is to present some material for the individual to study and later ask them to recall as much as possible without any further aids. This method is used in some tests for memory impairments; a short story is read to patients who then attempt to recall as much as they can. The first recall test is performed immediately after hearing the story, and usually a second recall attempt is asked for after a twenty- to thirty-minute delay. The differences between the two test results provides a measure of how rapidly meaningful material is forgotten.

In the laboratory, the experimental participants may be given a list of unconnected words to recall in any order they choose—a technique referred to as "free recall."

Using this method, researchers have discovered a lot about how memory works by varying the kinds of material used, rates of presentation, delays between presentation and test, ages of the participants, and many other variables. As we discussed earlier, it is often the case that information is still present in memory yet is difficult to retrieve. To address this problem of accessibility, cues can be given to remind the participants of unrecovered list words ("cued recall"), or the participant can be given a recognition test in which the original words to be learned are re-presented along with other words of a similar type; the task is to check off the original words.

Cued recall and recognition memory are more sensitive tests than free recall; typically recognition reveals greater evidence of memory than does cued recall, which in turn is better than free recall. When items are learned in the study phase, the encoded record of each event contains information about the item as well as its context, such as the size of print used, the item's color, where it was on the page, and so on. Surprisingly, sometimes people remember where the sought-after information appeared on the page but are unable to recall the information itself. As more information about the item and its context is supplied at the time of the test, the more likely it is that retrieval will be successful. The argument is that recognition supplies more relevant information than cued recall, which in turn supplies more than free recall.

Memory for Pictures and Words

One example of using different materials and different tests of memory performance comes from studies of pictures and words. Our memory for pictures is spectacularly good! In one experiment, the participants studied over six hundred pictures from magazines, and at the time of test were shown pairs of pictures, one of which had been studied, and the other of which was new. The participants had to choose the one they had studied; the average was almost 97 percent correct choices. In an even more dramatic demonstration of memory for pictures, psychologist Lionel Standing had experimental participants study ten thousand pictures for five seconds each over the course of several days![4] In a test session two days after this heroic learning experience, the subjects recognized 66 percent of the test pairs after the data had been corrected for guessing. Standing concluded that our memory for pictures is essentially unlimited.

Many studies have consistently shown better recall and recognition for pictures of objects compared to the object's name—such as a picture of a donkey compared to the word "DONKEY." This "picture superiority effect" is probably due partly to pictures providing more elaborate and distinctive features than their corresponding words, and partly to pictures later serving as more effective retrieval cues.

Memory for Names

One of the most aggravating aspects of our imperfect memories is the failure to recollect names of people we know as well as the names of animals, plants, cities, medications, and other proper nouns. Why are names so difficult to remember? In the case of surnames, one reason is that they are both arbitrary and lack meaningful associations. In the first place, there is no logical reason why the stranger on the subway car should be called Schmidt as opposed to Dupont or Anderson, and in the second place, the only associations a name may conjure up are other people with that name. For this second reason, uncommon names are particularly difficult to remember, and on trying to recall such a name, we typically resort to thoughts such as "I think it sounds like . . ." or "It probably has three syllables." Another point that makes names hard to retrieve is that they are highly *specific*. Unlike verbs, adjectives, and even common nouns, there is no way in which we can substitute a different phrase for the wanted name; we either get it or we don't!

The point that names have relatively few associated features has been illustrated in studies showing that the names of occupations are easier to remember than the names of people, even when it is the same word. So it has been demonstrated that the word "baker" presented in a list of occupations is better recalled later than the name

Baker presented in a list of names. This is likely because studying the word "baker" as an occupation brings to mind a rich number of thoughts such as that of baking bread, flour, an oven, and so on. In contrast, the name Baker brings little to mind except perhaps that it is a common name. This result leads to the thought that twisting a name to make something more meaningful of it will improve its later memorability. Of course, you have to be careful with this trick; if you make up a "memorable" name for a person that is humorous or mildly obscene, you may find yourself using it when you meet the person again!

Memory for High School Classmates

Studies have found that recognition memory for pictures is amazingly accurate. Suppose, however, you were shown a picture of your high school classmates with the names removed; you would probably fail to recognize some of the pictured people. On the other hand, it is likely that your performance would be higher than if the pictures were shown in a different context rather than with other high school classmates. Even people who are well-known are sometimes not recognized in an unusual context. A friend tells the story of traveling through Europe and on exiting a train, noticed a young woman who was also exiting the train but paid little attention to her. The young woman exclaimed, "Hi, Dad, did you not recognize me?!" He had not heard that she was traveling in Europe on holiday too. Even in a context that encourages recognition, we sometimes fail to recognize a known person because of a change in appearance. For older readers who have attended a high school reunion, you may have failed to recognize the once-shapely cheerleader who now has a very different shape, or the previous class nerd who is now well dressed and clearly prosperous.

In a large study conducted in the 1970s, researchers gave the participants a variety of tests for the names and faces of their high school graduating classmates.[5] The accuracy of the responses was verified from their high school yearbook. The participants had graduated at times ranging from three months to forty-seven years previously. One test was simply to recall the names of their classmates without any cues or reminders; understandably, success declined from 80 percent for the three-month group to less than 20 percent for the forty-seven-year group. Of course there are two factors at play here: the length of time since graduating, and the fact that those graduating forty-seven years earlier were necessarily older. It is well established that memory is poorer at longer delays and also in older people (see chapter 6).

A further finding from the study was that when the names and faces of former classmates were both provided, the ability to match the names and faces remained steady at approximately 90 percent correct for at least fifteen years, even for members of big classes. It is important to bear in mind that the participants in this study knew that they were trying to recall and recognize names and faces from their high school. If a pictured person was shown in isolation without this information, name recall and even face recognition would probably be much lower. And if you are asked to recall as many classmates as you can, you would likely think back to your high school days as a means of gaining context that would prompt and support your memory. As we discuss later in the chapter, the reinstatement of an event's original context is crucial for effective recollection.

In a further study, the participants were asked to think aloud while trying to recall the names of their high school classmates. The protocols resulting from their thinking aloud provided evidence that the task of recalling names was treated as a type of problem-solving. The process of remembering typically began by the person thinking back to find a context that provided an environment in which a search for a targeted name could be conducted. This process of establishing the details of the original situation went on for a surprisingly long time, and included hypothesizing, establishment of search contexts, use of several search strategies, and self-correction. More generally, this type of remembering is best thought of as a problem-solving process involving a reconstruction of the past.

The goal of current perception is important for what is later remembered. For example, a study was conducted in which some participants were led through a house after being instructed to adopt the perspective of a potential burglar, whereas another group went through the same house adopting the perspective of a home buyer.[6] Not surprisingly, the perspective adopted influenced what was best remembered, with those taking the perspective of a burglar best remembering objects attractive for theft, whereas those taking the perspective of a home buyer best remembering aspects of the house relevant for its purchase. Remembering, whether for a name or the contents of a house, relies on reconstructive processes, which in turn are heavily influenced by the purposes and goals of the original perception.

Feeling of Knowing

When trying to recall the name of an acquaintance or a former high school classmate on seeing their photograph, you may feel certain that you *know* the name although you can't recall it at the moment. This pervasive "feeling of knowing" was first studied by J. T. Hart in the 1960s. The participants were given general knowledge questions such as, "Who was the first person to step on the moon?" For questions answered incorrectly or for which an answer

The goal of current perception is important for what is later remembered.

could not be produced, people were asked if they would recognize the correct answer in a multiple-choice recognition test. Recognition memory was much more accurate when they predicted that they would recognize the correct answer than when they predicted they would not. Many subsequent studies have shown that feeling-of-knowing judgments are generally accurate, although sometimes they may be greatly in error. In chapter 4, we describe reasons for such misplaced high confidence in incorrect responses as well as other examples of false memory.

The finding that feeling-of-knowing judgments are generally accurate provides evidence that there is much information in our long-term memory store that we are unable to recall although we somehow know it is there. We can retrieve some of the item's features but not its name or perhaps fail to recollect the situation in which the event occurred. The tip-of-the-tongue phenomenon is a subjective state of this sort; it provides further evidence for the success of recognition memory when recall fails.

The Tip-of-the-Tongue Experience

Have you ever had the experience of attempting to recall the name of a person, place, or thing, were unable to do so, but had the agonizing feeling that the name is "on the tip of your tongue"? Everybody responds "yes" to this question, and if you are an older adult, you will likely say that it happens often. The tip-of-the-tongue phenomenon (usually abbreviated to TOT) was first investigated by the US psychologists Roger Brown and David McNeill in the 1960s.[7] They gave people definitions of uncommon words and asked them to produce the defined word. One example was, "What is the name of the navigational instrument used in measuring angular distances, especially the altitude of the sun, moon, and stars at sea?" For some of the questions asked, the participants in the TOT state were unable to produce the defined word but reported feeling certain that they knew the answer. When in that state, they appeared to be in mild torment, much as if they were preparing to sneeze. On further questioning, it was found that they frequently knew the initial letter of the sought-after word, less often knew the last letter, and even less often knew the middle letters. The number of syllables was often known, and the participants expressed confidence that they would recognize the word as the one they were searching for if it was provided. Does all of this sound familiar? On the off chance that the reader is in the TOT state trying to recall the word defined above, the word is "sextant."

In some studies, people were asked to record the details of TOT states as they occurred day by day. These diary studies have shown that TOT states occur most for proper names and names of objects, and that older adults experience TOT states more frequently than do young

Figure 3 Tip of the tongue in action! *Speed Bump* cartoon reproduced by permission of Creators Syndicate Inc.

adults. When a person is in a TOT state, a common strategy is to think back to a context in which the name was encountered. An alternative strategy is to generate words from the presumed knowledge of the name's first letter or speech sound. Reliance on context has generally been shown to be the more effective of the two strategies. The typical rate of successfully resolving a TOT state is over 90 percent, based on reports from people who have been asked to record their daily experiences. The sought-for word or name is typically found or spontaneously comes to mind shortly after one has given up on the search. The name sometimes "pops" to mind much later, though, long after abandoning the attempt. When this happens, it can be so surprising that you say the name aloud, puzzling those around you! It seems likely that this later retrieval success is triggered by hearing or seeing the word itself, or encountering a word or object that is closely related to the missing word.

The TOT phenomenon provides evidence to support models that distinguish among different forms of mental representation of words and objects. One form of representation is visual—what the target object or named person looks like. A second aspect is a conceptual representation that specifies what functions an object performs or biographical facts about a person. Third, a phonological representation specifies the name's constituent sounds. A deficit in memory can be specific to one or more of these different forms of representation. For the TOT phenomenon, however, it appears to be the phonological representation that is not fully accessed. Some researchers have suggested that the neural linkages between a word's meaning and speech sound representations are temporarily disconnected in the TOT state.

Context Effects

In the previous sections, the reconstruction of the original context for a wanted name or event is carried out intentionally. Yet sometimes an aspect of the current context can bring memories back to mind spontaneously. For example, on returning to an area in which you lived many years ago, you may be flooded with memories from the past. Also, aspects of your behavior might change in surprising ways. If you were raised in the South and moved to the North, you might find your accent becoming more southern on returning home, surrounded by your old friends. In this vein, one of the authors once went to a pseudo-Scottish pub in Toronto with a friend (the *other* author, actually!) who was originally from Scotland. On entering the pub, his accent became sufficiently Scottish to make it difficult to understand what he said!

Such effects of context change have been revealed in experiments. For example, two British psychologists whose hobby was diving examined memory for words presented auditorily (by telephone cable) to divers while underwater, with memory being tested while they were either again underwater or on land.[8] For a *recall* test, words learned underwater were best recalled (by writing the words on a waterproof pad) when again tested underwater. In contrast, recognition memory performance did not differ as a result of a change of context between the study and test. As described earlier, recognition memory is largely independent of context changes whereas the processes of recall are highly reconstructive, and being back in the learning context helps such reconstructive efforts.

The point that context changes can have large effects on memory has been shown in investigations of *state-dependent memory*. That is, an event occurring when a person is in a given mental state (e.g., elated, depressed, or intoxicated) is best recalled when the person is again in that original state. There's the tale of a relative cleaning an alcoholic's house after his death and finding liquor bottles hidden in many places. The suggested reason was not that he always wanted to have a drink nearby but rather that he hid the bottles while intoxicated and could not find them when sober. If he could only have another drink, he might have located a missing bottle, but . . . you see the problem! Laboratory studies have provided more formal proofs of such state-dependent effects of intoxication.[9]

An event occurring when a person is in a given mental state is best recalled when the person is again in that original state.

Many experiments have been done to show the importance of mood congruity between study and test. For such studies, mood may be induced by reading stories that make the participant either happy or sad. Studies have also been carried out on patients who are prone to depressive episodes, and having them learn and recall material either in the same state (calm or depressed) or the opposite state during learning and retention. The results from such studies are much the same as described earlier for the effects of intoxication. Eric Eich provides a review of the results from investigations of state-dependent memory.[10] Included in that review is a case of a person with multiple personalities, with the results showing that memory relied on the match between the personality that obtained a piece of information and the reinstatement of that personality during the test. Less extreme forms of a result of this sort are likely common. For example, one's personality at work might differ from that at home, with corresponding effects on memory. To an extent, we all have multiple personalities. In general, the effects of context change or reinstatement are central to many memory successes and failures. To repeat an important point in this regard, the beneficial effects of being in the same mental or physical context are greater for unaided recall than for recognition memory, with the reason being that reconstructive processes are more dominant when you have to bootstrap the processes of remembering by attempting to recollect aspects of the original event.

As a further illustration of context effects, studies have shown that students do better on a test if tested in the same room as they studied.[11] If told to imagine that they are back in the room in which they studied prior to completing the test, however, the effect of the room change disappears. This occurs only if the material studied and tested is not difficult; for more difficult material, greater reconstructive processing is likely required, and that is aided by details of the original context. One of us was giving a test and noticed several students staring intently at him. What was the problem? Were they beaming their indignation because the test was too difficult? Instead, it is likely they were trying to consciously re-create experiences associated with hearing the original relevant lecture.

The Effects of Spacing Repetitions and Distribution of Practice

What is the most efficient way of learning new information or a new skill? The most obvious answer is to keep repeating the appropriate learning operations until success is achieved. There is the old story of the visitor to New York

City asking a local citizen, "How do you get to Carnegie Hall?" to which the local, a professional musician, replies, "Practice, practice, practice!" Perhaps surprisingly, though, simple repetition is not always the best method. In chapter 2, we described an experiment showing that the deep semantic processing of words is much more beneficial for later memory than is the "shallow" processing of rhymes or typescript. Related work has found that the rote repetition of words is not an effective method for later recall.

One important method for improving the effectiveness of repetitions is to space them out rather than repeat them in a "massed" sequence. This is especially true when other items to be learned intervene between repetitions of the item we are focusing on. This "spacing effect" is robust, and easily demonstrated using either recall or recognition memory. The reasons behind the effect are still being debated. One suggestion is that if a word to be learned is repeated immediately after its first presentation, it does not receive much attention; the experimental participant regards it as already learned. If the word is repeated after some time, however, it may be recognized as a possible repetition and attract some attention to assess that possibility; at longer intervals, such assessments are increasingly effortful, and this effortful processing is helpful for later recall. Another idea is that if the item or event to be learned is repeated at a later time, it will also be in a slightly different context, and this contextual variability enriches the item's encoded representation, making it more memorable.

Turning to more applied settings, many studies have been done to show the advantages of distributed practice over massed practice. For example, a study by the British researchers A. D. Baddeley and D. J. A. Longman in 1978 involved the training of postal workers to perform a keyboarding skill. The study demonstrated that spacing the learning sessions over days produced better learning than massed practice in a single day. Distributed practice has been shown to be advantageous for a wide variety of skills, including playing a musical instrument, attaining language skills, and putting in golf.

An impressive example of the advantages of distributed practice for learning a foreign language is provided in an article by Harry Bahrick, Lorraine Bahrick, Audrey Bahrick, and Phyllis Bahrick.[12] Theirs was a nine-year longitudinal study in which four participants learned three hundred English–non-English-language word pairs. There were either thirteen or twenty-six relearning sessions that occurred over intervals of fourteen, twenty-eight, or fifty-six days. Retention was tested for one, two, three, or five years after the training terminated. The tests of retention revealed that the fifty-six-day interval between the training sessions produced better retention than did the twenty-eight-day interval, which in turn produced higher retention than did the fourteen-day interval. Also, the rate of forgetting across years was slower for the fifty-six-day interval between the training. The reader might ask how the participants were recruited for engagement in such a demanding study. A clue is provided by the fact that the authors all have the same surname!

Repetition of similar events happens all the time in our everyday lives, giving rise to another aspect of memory: Which event occurred more recently? Suppose you were asked which of two movies you saw more recently. You would likely be more accurate if viewing the later movie had reminded you of the earlier movie. The US psychologist Douglas Hintzman has suggested that this kind of re-minding occurs frequently as we go through life, and that the resulting process provides a framework for remembering the order in which events occur in our lives—a kind of mental time line.[13] How people remember the relative order of events is the focus of many interesting experimental studies.[14]

Retrieval as Learning

A number of studies have shown that retrieving information from long-term memory is a surprisingly effective learning method. One account of this effect is that retrieving the item typically involves thinking about its meaningful features so that retrieval processes are very much like the deeper encoding processes in the levels-of-processing framework described in chapter 2. If the second presentation of the word to be learned in the spaced-learning situation (described in the previous section) is delayed long enough, its recognition as a repeated item will also involve deeper retrieval processing, and this may be a further reason for the beneficial effects of spacing.

The advantages of retrieval for subsequent memory are also evidenced by the effects of testing. Henry L. Roediger III and Jeffrey D. Karpicke conducted an experiment in which students studied a descriptive passage about science, and were then either given a second learning opportunity or a retrieval test of memory for the passage.[15] The memory test was given after five minutes, two days, or one week, and the two groups were then given a final test of recall of the original passage. After five minutes, restudying produced better recall than did a retrieval test. Yet after both two days and one week, prior testing produced a much better final recall performance than did restudying. Clearly, giving quiz tests can result in better subsequent learning than a further study session. The results, though, are probably dependent on the nature of the material and form of the test. For example, with a sufficiently difficult text, rereading is required for comprehension. Also, when reading a difficult text, a student will often engage in self-initiated retrieval with retrieval aimed at dealing with difficulties in comprehension.

Retrieving information from long-term memory is a surprisingly effective learning method.

Returning to the problem of remembering names, Thomas K. Landauer and Robert A. Bjork suggested that after meeting a new person, covert retrieval of the person's name is a good way to learn it.[16] They also showed that gradually increasing the spacing of retrievals is an even better method to retain the name in memory. To use that technique, during a conversation with a new acquaintance one should first repeat their name immediately after being introduced, work the name into the conversation again a rather short while later, and yet again after longer intervals. To be most effective, the intervals between repetitions should be such that retrieval is difficult but still possible. Of course, if you go to extremes using this method, the person you just met will find you a little creepy and may avoid you in the future, giving you no reason to remember the name! It is probably better to silently retrieve the name or retrieve the name when talking about the person with others. It is, however, a remarkably effective way to learn new names.

FORMS OF MEMORY

Processes, Dissociations, and Attributions

We normally think of memories as conscious experiences—either as recollections of past events or facts we have just retrieved. Yet good evidence now shows that past experiences can affect our behavior in the absence of conscious awareness. In earlier chapters, we described such forms of memory in amnesic patients. For example, patient H. M. could perform a skilled motor task while having no recollection of practicing it earlier; another deeply amnesic patient did not laugh when he was told a joke for the second time, although he had no conscious memory of hearing it before.

In this chapter, we look at corresponding cases of memory without awareness in both young and older adults with normal memory. These forms of memory without awareness are often associated with a decline in cognitive

Past experiences can affect our behavior in the absence of conscious awareness.

control and instead show a reliance on a more automatic, habit-like form of memory.

Such habitual types of memory frequently reflect many previous occurrences and past learning opportunities. They are generally helpful and adaptive in the sense that they enable us to perform such recurring actions as turning off light switches or locking doors without the involvement of controlled attention and conscious decision-making. The downside is that such habitual and near-automatic actions leave sparse records in conscious memory so that individuals often worry later, "Did I turn off the stove?" or "Did I remember to lock the door?" As we have just observed, these unconscious habitual influences are usually shaped by many prior instances, but recent research has found that a single event can influence later cognitive processing in much the same way. We examine such cases in the following section.

Effects of a Prior Exposure on Later Performance

The example of a joke being spoiled for an amnesiac by its prior telling shows that a single event can produce a habit-like unconscious influence of memory. More formal evidence of that sort has been found for amnesiacs. In one such experiment, English psychologists Elizabeth K. Warrington and Lawrence Weiskrantz presented a list of five-letter words to be read by amnesic patients and healthy control participants, and then assessed memory for the earlier-read words after a delay, using two different tests.[1] In a word completion test using the first two or three letters of each word as cues, amnesiacs were as likely as participants with normal memory to produce earlier-read words. All the participants were then given a recognition memory test for words they had read earlier. For the participants with normal memory, performance on the recognition memory test was high but recognition was at the level of chance for amnesiacs. Although the patients did not recollect studying the list, the recently presented words strongly came to mind in the word completion test and were given as solutions to the relevant fragments. This result for amnesiacs amounts to a dissociation between an unaware use of memory (what comes to mind) and aware use of memory (recognition memory).

An experiment carried out in the early 1980s by Larry Jacoby and Mark Dallas produced results showing a dissociation between aware and unaware influences for people with normal memory similar to that found for amnesiacs.[2] Results from their study showed that a single prior experience can influence what is subsequently seen even though the particular prior experience is itself not remembered. In this study, college undergraduates first answered questions about words presented in a long list. The questions were either about a constituent letter (e.g., Does the word contain the letter *B*?), rhyme (Does the word rhyme with "TRAIN"?), or the word's meaning (Is the word a piece of furniture?). This is the "levels of processing" manipulation described in chapter 2, and the typical finding is that meaning is better than letter or rhyme questions for later conscious memory (see figure 2). This result was again obtained in the present study for one group of participants who were given a recognition test after answering the questions posed in the first list.

A second group of participants learned the word list in the same way, but were given a test in which words were flashed for a brief duration on a monitor; the task was simply to identify the flashed word if they could. This test of *perceptual identification* included some words from the learned list, but the participants were not informed of this. The finding was that words from the first list were identified more often than completely new words, but in contrast to conscious recognition memory, the levels of processing manipulation (letter, rhyme, or meaning questions in the first list) did not change the performance levels. Further work showed that a flashed word was just as likely to be identified when it was *not* later recognized as when it *was* recognized in a later test. It was also found that the effect of reading words in the experimental setting was long lasting, continuing to be present several days after the initial reading, although recognition memory performance declined across that interval. If the first-list words were presented auditorily, however, these words did *not* increase later visual perceptual identification, demonstrating that the effects of this type of unconscious memory are quite specific to the way that the word was initially presented.

Note that the results reveal the same pattern as found for amnesiacs, showing memory for a particular prior experience without being aware of the prior experience. Others later used word completion tasks of the sort used by Warrington and Weiskrantz, described above, to obtain results like those reported by Jacoby and Dallas. Such results can be characterized as relying on an automatic influence of memory. In a further demonstration of the same basic phenomenon, perceptual psychologist Paul Kolers had the participants practice reading text passages in transformed typographies (for example, the words were shown as inverted mirror images).[3] Early in such practice, words from the text were well recognized in a subsequent memory test. As the training progressed, however, and reading gradually became faster and more fluent, the accuracy of later recognition memory for words in each novel text declined to the point of being near zero. The consciously controlled reading of individual words was reduced, being replaced by a more skillful, automatic reading of the text, with the result that reading each word became more like the automatic action of switching off a light. In a further experiment, Kolers had the participants reread passages in a transformed typescript that they had first read a year earlier, along with some new passages. The participants reread the original passages faster than the new ones, although in many instances the readers did not recognize the passages as ones they had read before. Kolers therefore concluded that "knowledge can be expressed as skilled performance without a corresponding recognition in conscious judgment." This finding is similar to the observations on patient H. M. described in chapter 1, although in the present case the participants were college students with normal memories.

Following the publication of the Jacoby and Dallas paper, several studies were done using a word-completion task similar to that used by Warrington and Weiskrantz. These studies revealed a dissociation between performance on a word-completion task and recognition memory performance for the participants with normal memory, just as previously found for amnesiacs. Daniel Schacter provided a review of such experiments.[4] As described in the following sections, there are advantages to be gained by separating forms of processes within a task rather than identifying forms of memory with different tasks. That is, tasks are not process pure, in that the performance of a given task might rely on different processes in different situations.

Recognition Memory and Memory Attributions

Jacoby and Dallas related their results to dual process accounts of recognition memory: the idea that we can recognize a person either by consciously recollecting the circumstances under which we met the person previously or more simply by the feeling of familiarity that seeing the person engenders. US psychologist George Mandler illustrated this distinction by "the butcher on the bus" example: the experience most people have had of encountering a known person in unusual surroundings, like an office coworker in a weekend supermarket or indeed spotting your butcher on a bus.[5] Such meetings often give rise to strong feelings of familiarity in the absence of conscious recollection of how you know the person. Jacoby and Dallas noted that this feeling of familiarity is sometimes misleading in that it reflects a *process of attribution*. That is, the feeling of familiarity is interpreted by the cognitive system as reflecting previously acquired knowledge, but this may not always be the case. For instance, rather than arising from a prior encounter with the butcher, the feeling of familiarity might have arisen from having briefly seen the person among those in the crowd waiting to board the bus.

The proposal is that we possess a sense of how easy or difficult it ought to be to perceive any object, person, picture, or melody, and if the perceptual processing is easier or more fluent than it should be, we attribute this enhanced fluency to memory familiarity—"we must know this person from some previous occasion." The complementary process of *conscious recollection* is much less automatic; it depends on effortful cognitive control processes that rely on the retrieval of the context or circumstances under which we know the person. As a related example of misattribution that reflects fluency, a previous hearing of a melody can increase a person's later liking of the melody even if the person does not recognize it as one that was heard earlier. In that case, familiarity is misattributed to the inherent pleasantness of the melody. Studies have shown such effects on the likability of melodies. In what follows, we describe false memories that arise from misattributions of fluent processing.

Such misattributions have been used to explain the déjà vu experience of falsely having previously experienced an event. In 1928, psychologist Edward Titchener described false recognition as an illusion of memory that is produced by "a disjunction of processes that are normally held together in a conscious present."[6] He illustrated his argument with the example of a person who hastily glances across a street in preparation for crossing and is then momentarily distracted by the contents of a store window. On crossing the street, the person experiences false recognition as the feeling of having previously crossed that same street—a feeling of déjà vu. By Titchener's account, "the preliminary glance, which naturally connects with the crossing in a single, total experience, is disjoined from the crossing, and comes to consciousness separately as the memory of a previous passage." This is portrayed as the severing of "two phases of a single consciousness; the one is referred to the past; and the other, under the regular laws of memory, arouses the feeling of familiarity."

In a similar way, the effects of a fleeting glance can be misattributed in a way that gives rise to false remembering. An experiment was designed to gain further evidence of this sort.[7] In a first phase, a long list of words was presented for study. In a subsequent visual recognition memory test, previously presented and new words were sometimes preceded by a flashed word (a "context word") that was identical to the test word. When the presentation of the preceding word was so brief that it was not consciously perceived, its presentation increased the probability of mistakenly judging a new test word to be one studied earlier. The researchers argued that the unconscious perception of the context word increased the perceived fluency of reading the test word. In turn, this increased fluency was misattributed to familiarity stemming from prior study of the test word. In contrast, if the context word was presented for a duration sufficiently long for it to be consciously perceived, the opposite effect was found: its presentation *decreased* the probability of a new word being mistakenly judged to have come from the studied list. In this second case, any familiarity of a new test word was attributed to its immediate prior presentation rather than to its presence on the original list.

Experiments have shown that illusory familiarity can also be produced by varying the perceptual clarity of a target. For example, Bruce Whittlesea had the participants study a list of words presented briefly (less than one-tenth of a second for each word).[8] In a later recognition memory test, the ease of reading the words was varied by presenting them visually under superimposed random patterns that were either light (so that the word could be read fairly easily) or heavy (making the word difficult to read). The participants were instructed to first identify the words, and then judge them as old (previously seen) or new. The results revealed that both old and new words were more likely to be judged as old when they were lightly masked by the random pattern. That is, the ease of reading the lightly masked words was misattributed to their prior presentation. Other investigators have demonstrated a similar illusion of familiarity by briefly presenting a list of spoken words, and then testing for recognition in the context of a soft or loud background noise. Words tested with a soft background noise were more likely to be called "old."

Misattributions of familiarity are not restricted to perceptual fluency but can also arise from conceptual fluency (the ease with which a category member comes to mind) and retrieval fluency (the ease with which a sought-for response comes to mind). As a commonplace example of conceptual fluency, readers may have had the experience of telling a joke to the very person from whom they first heard it. On meeting that person, the joke readily comes to mind, is thought of as one the person would like, and then retold—followed by embarrassment when their mistake is exposed!

Misattributions of the past can result in perceptual illusions too. For instance, Dawn Witherspoon and Lorraine Allan had the participants study a list of words prior to judging how long briefly presented words remained on the screen.[9] The results showed that earlier-read words were judged as staying on the screen longer than were new ones. A parallel effect has been found for judgments of background noise paired with spoken sentences. The background noise accompanying sentences that had been presented previously was judged to be less loud than the noise accompanying sentences that had not been heard earlier. This effect is striking in that it does not disappear when the participants are informed of its cause.

Colleen Kelley and Larry Jacoby carried out a study showing that variations of relative retrieval fluency can result in a misattribution of difficulty for others.[10] The task was to solve anagrams, and in one condition the participants had previously read a list of words containing solutions to the anagrams, although they were not

Misattributions of the past can result in perceptual illusions.

informed of this fact. The presence of solution words in the first phase led to the faster solution of anagrams presented in the second phase. When the participants were asked to judge the difficulty of the anagrams for others, it was found that the anagrams whose solution had been presented previously were judged to be easier for others to solve than were anagrams with no previous solutions. That is, the ease of solving an anagram was misattributed to the structure of the anagram itself rather than to an earlier reading of the solution word. This illusion of the ease of solving by others was removed when the solution word was presented *with* the anagram, making the relationship between the solution word and anagram obvious. These results thus show that people may fail to recognize that their personal subjective experience of difficulty may not generalize to others—a form of "adult egocentrism." Illusions of this sort often occur when an expert misjudges a task as being easy for a novice. For example, it is common for teachers to underestimate learners' difficulty in understanding material. An excellent review of the role of memory attributions in memory and judgments is provided by Colleen Kelley and Matthew Rhodes.[11]

Results of the sort described above demonstrate that we have much in common with amnesic patients in our showing of unaware influences in memory in combination with attribution processes. Our past and present are frequently inferred, sometimes incorrectly, instead of being

People may fail to recognize that their personal subjective experience of difficulty may not generalize to others—a form of "adult egocentrism."

remembered or directly experienced. Such results were evidenced by showing effects on a test that revealed the effects of memory that did not rely on an awareness of the relevant memory. Much can be gained, however, by separating the different forms or uses of memory within a task as opposed to identifying them with separate tasks.

Separating the Contributions of Familiarity and Recollection

Is *Sebastian Weisdorf* famous? He was not until he was made famous in an experiment conducted by Jacoby and associates.[12] The basic idea behind the study is that many cognitive processes are affected by both conscious and unconscious influences associated with past experiences; one goal was to provide separate measures of these influences. The participants in the experiment read a list of made-up fictitious names, including Sebastian Weisdorf. They were correctly told that the names were fictitious and therefore not the names of famous people. Next, the participants were presented with a second list of names, including some from the first list mixed up with names of actual famous individuals and new fictitious names— ones not earlier presented. The task was to judge whether each name was of a famous person. The participants were warned that the list contained names presented in the earlier list and reminded that the names included in the earlier list were all nonfamous. This test occurred either immediately after reading the list of nonfamous names or twenty-four hours later. In terms of the processes discussed in the present chapter, the feeling of familiarity associated with previously read fictitious names would suggest that the name was of a *famous* individual, whereas if the participant recollected that the name was on the first list, the processes of cognitive control would oppose that judgment. The ability to recollect an earlier reading of a name in the list of nonfamous names was expected to decline across time, being higher for the immediate test than for the test delayed by twenty-four hours. Indeed, results from the immediate test showed that the probability of mistakenly accepting an earlier-read nonfamous name was lower than that for accepting a new nonfamous name. In contrast, the opposite was found when the test was delayed for twenty-four hours; the participants wrongly judged previously read nonfamous names to be famous much more often than they did new nonfamous names, showing that the influence of familiarity was less likely to be successfully opposed by recollection. The finding that fictitious names mistakenly became famous as a result of being earlier read serves as an example of the misattribution of familiarity. The finding fits well with the politician's line, "I don't care what they say about me so long as they spell my name correctly!" Mere repetition can increase the illusion of fame (unless of course the listener recollects the many negative things that the politician may have done in the past!).

The magnitude of the false fame effect depends on the failure of recollection; it was shown to be larger after a twenty-four-hour delay than when the fame test was given immediately after reading the first list of fictitious names. The effect is also larger when the participant's attention is divided while judging fame, and when the participants are required to respond rapidly on the fame judgment test—manipulations that again reduce the probability of recollection. Older adults are more prone to the attribution of false fame than are younger adults, suggesting an age-related decline in cognitive control and thus the ability to recollect. Finally, the false fame effect has been found in amnesic patients too. Interestingly, the effect is even larger in patients than in healthy adults, in line with the notion that although amnesiacs' ability to recollect events is impaired, their unconscious habitual memory processes are intact, and exert a relatively greater influence on thought and behavior.

"Did I tell you this story before?" As one becomes older, such questions are asked more frequently in an attempt to avoid unwanted repetitions due to a decline in recollection as a means of opposing the automatic retelling of a story. An experiment by Janine Jennings and Larry Jacoby showed that older adults were less able to avoid repetitions than were young adults.[13] Also, it is likely that oft-told tales about the past, such as family legends, do not truly reflect recollection of a prior event but instead result from a more automatic, habit-like influence of memory. Memory is likely to have been shaped and elaborated across the many retellings of a story to produce a better story, more suitable for a general audience. If you listen carefully to yet another retelling of a friend's favorite story, you are likely to find that the current version is a near-verbatim repetition of an earlier version as opposed to a recollection of the original event.

Measuring Automatic and Controlled Influences: Process Dissociation

An elderly professor in Winnipeg, Canada, traveled to a conference in Chicago but was unable to find his airline ticket when returning home. After an extended search for his ticket failed, he bought a return ticket and called his wife to pick him up at the airport on his arrival home. His wife responded that she would love to do so, but they had only one car and he had driven it to the conference!

The hapless professor was in a situation in which the appropriate response was opposed by a more habitual, automatic response—the response of flying home from a conference as he typically did. His doing so is akin to mistakenly judging an earlier-read fictional name to be famous, inappropriately relying on familiarity—an automatic influence of memory. Having driven to the conference, habit led him to buy a return ticket, whereas recollection would have opposed his doing so. In contrast, had he flown to the conference as he usually did and then lost his ticket, habit and recollection would have acted in concert to result in his correctly buying a replacement ticket. The accuracy of responding when habit and recollection act in concert can be contrasted to accuracy when habit and recollection act in opposition (with habit suggesting one response and recollection suggesting the opposite response). The influence of recollection can then be measured as the difference in performance between the "in concert" versus "opposition" condition. In the case of the errant professor, if he was as likely to buy a return airline ticket after driving to the conference as after flying to the conference, it could be concluded that he was fully incapable of recollecting his initial means of transportation; his estimated recollection would be zero (a probability of 0), and his reliance on the automatic influences of memory (habit) would be total (a probability of 1.0).

Janine Hay and Larry Jacoby used an associative memory task to mimic the above example.[14] During a training phase of the experiment, habit was manipulated by presenting pairs of words to be read, varying the probability of one of two right-hand words being associated with the same left-hand word. For example, the pair "knee bone" was presented on 75 percent of the training trials, whereas the pair "knee bend" was presented on the remaining 25 percent, making "bone" the typical response (akin to typically flying home from a conference).

Following the training phase, a list of pairs was presented to be studied for a following memory test. For an in concert condition, some pairs were those made typical by prior training (e.g., knee bone). For that condition, a later correct recall of the right-hand member of the pair (bone) given the left-hand member (knee) as a cue could result either from the typical habit created during training or a recollection of it having been paired with the left-hand member during study. In the opposition condition, the pair presented for study was the atypical pair (e.g., knee bend). For that condition, recollection would produce a correct response (bend), whereas relying on habit would result in an error, producing the word made typical during training (bone). Results from the in concert and opposition conditions were combined to estimate the probability of recollection as well as the probability of relying on habit. The experiment included both young and elderly participants. The results revealed that estimated recollection was much lower for older adults than for young ones (0.29 versus 0.44), but the estimated influence of habit was identical in the two age groups (0.72 versus 0.72).

Judy Caldwell and Michael Masson used a version of this *process dissociation procedure* and also found that older adults had a deficit in the probability of recollection, but showed no age-related difference in the automatic influences of memory.[15] Their experiments tested memory for object locations by having the participants search for objects. The results from their experiment are in line with reports from older adults who often have difficulty finding household objects, typically looking for them where they are usually found rather than recollecting where they were last left.

David Balota from Washington University in Saint Louis performed a study with his associates to show that a version of the process dissociation procedure was useful for identifying people in an early stage of Alzheimer's disease.[16] In that study, the probability of recollection was found to decline successively from middle-aged adults to older ones to those in the early stages of Alzheimer's disease. In contrast, the estimate of automatic influences was found to be invariant across the different groups of participants. As will be described in chapter 6, deficits in memory in older adults as well as those in Alzheimer's patients commonly involve a deficit in cognitive control resulting in failures of recollection.

Estimates of recollection gained from the process dissociation procedure are often equivalent to measures based on the subjective experience of remembering, but not in all cases. The notion of "remember/know" judgments, proposed by Endel Tulving and used extensively by our colleague John Gardiner, provides a measure of the phenomenological experience of memory.[17] The participants are told to say "remember" if the recognition memory of an event is accompanied by details of its time and place of occurrence, and thoughts about the event when it was experienced. A "know" response was used to indicate a feeling of familiarity or certainty that the event had been experienced earlier, but without any details of its prior occurrence. (The puzzled person on the bus might *know* that she had encountered the butcher somewhere, but have no idea about who the butcher was, or where or when they had met. Her friend, on the other hand, might *remember* that the person is the butcher from the local supermarket and that she had bought steak from him yesterday.)

Additionally, in this technique, the participants were instructed to say "guess" if their judgments of the remembered items were based on little evidence. Early applications of the remember/know procedure found that older adults "remember" less but "know" more than do young adults—a conclusion that the elderly authors of this book quite like! Yet given that there are three categories for judgments, the increase in "remember" responses found in young adults necessarily reduces the possibility of "know" and "guess" replies. To take this restriction into account, the probability of a "know" response can be corrected statistically.[18] Rather sadly, doing so results in the conclusion that older adults remember less, but do not differ from young adults in the amount that they know. On the positive side, however, this conclusion is in line with findings from the process dissociation procedure.

Proactive Effects of Memory: Interference versus Facilitation

The elderly professor's error of buying an airline ticket to return home can be described as an example of proactive interference of the sort investigated in the verbal learning tradition described in chapter 1. There, proactive interference was illustrated by the difficulty of learning Spanish after learning French as a second language; the French words keep intruding when you attempt to retrieve the Spanish words. Similarly, the error made by the professor resulted from a mistaken reliance on the past.

More recent investigations by Christopher Wahlheim and Larry Jacoby have shown the critical importance of *noticing* and *recollection of change* for understanding the effects of proactive influences on memory.[19] In the case of the elderly professor, he undoubtedly noticed the change from flying when he drove to the conference. Yet he failed to recollect that change when preparing to return home, resulting in his buying a return ticket, mistakenly responding to a habit gained from the past. What Wahlheim and Jacoby found is that if people notice the change initially and recollect the original response (while being aware that it is no longer the correct one), this can transform proactive interference into proactive facilitation. That is, the correct recall of the original association can facilitate the use of the new association. Proactive facilitation results because noticing change integrates the earlier response in a memory representation that includes the later response along with their order of occurrence. A recollection of change at the time of the test results in memory for the earlier response bringing the changed response to mind, facilitating its recall. If the change is noticed but not recollected, proactive interference results to an extent that depends on the greater strength of the earlier response.

Noticing change influences perception as well as memory. As an example, Alinda Friedman showed participants pictures of items in a common or uncommon context.[20] One picture, for instance, showed a toaster in a kitchen, whereas another showed a horse in a kitchen; the inappropriately changed context is obvious in the second case. The results demonstrated that eye movements directed at the horse were much more frequent than were those directed at the toaster, and subsequent recognition memory for having seen the horse was much higher than recognition of the toaster. In contrast, Bill Brewer provided evidence that expectations from the past, which can be thought of as being a form of proactive interference, can result in false memory.[21] The participants in his study viewed an office and later reported memory for its contents. Memory was excellent for objects that are typically seen in an office such as a desk and chair, although the participants spent less time looking at those objects. More important, many of the participants reported seeing books in the office although no books had been there—an unusual change from normal circumstances. Similarly, errors resulting from a reliance on a schema to reconstruct the past of the sort found by Bartlett (chapter 1) can be described as resulting from proactive interference. In other cases, a reliance on the reconstruction of the past facilitates memory, as noted in chapter 3 for remembering the names of high school classmates.

Metacognition

The term "metacognition" refers to thoughts and beliefs about our own cognitive processes, including memory. In chapter 3, we presented evidence to show that a recall test often does more than further studying to enhance later memory. Students, however, commonly report the erroneous belief that rereading is the better method for study. Also, students are often in error when judging their own level of learning—another measure of metacognition that has been frequently investigated. Such subjective impressions are important for determining the methods used and lengths of time that students devote to studying. Good students know best how to study and are better at judging their level of learning, and this influences how much they study. On the other hand, poor students after taking a test and prior to obtaining the results frequently predict that their grade on the test will be much higher than that obtained. This poor assessment of their own ability is a form of "unawareness" that points to their poor metacognition.

Metacognitive thoughts are common and have been described in earlier chapters. For example, the feeling of knowing is an aspect of metacognition, as are thoughts involved in the tip-of-the-tongue experience, and the feeling of high confidence that sometimes accompanies a memory response. Such awareness of memory processes and their limitations is crucial for guiding behavior. Both younger children and older adults typically fail to make spontaneous use of learning strategies that they are capable of using when instructed to do so. The term *production deficiency* has been used to portray this state of affairs. This processing inefficiency is particularly true for older adults, who often fail to engage in a cognitively controlled form of remembering, recollection, but instead are more prone to errors resulting from an overreliance on automatic influences of memory. This parallel between children and young adults is understandable when we bear in mind that the use of strategies is a function of the brain's frontal lobes. As will be discussed in chapter 7, frontal lobe functions involved in cognitive control mature during childhood, adolescence, and early adulthood, and decline in old age.

CREATING MEMORIES
True and False

Emotion and Memory

Emotional events have a profound impact on memory. We have vivid memories of both the triumphs and tragedies in our lives. For instance, when identifying the perpetrator, a rape victim might proclaim that they will never forget the person's face, saying that it is indelibly represented in their memory. Such testimony is compelling, but is the memory an accurate one? Not always, as it turns out. A striking instance of memory failure in such a situation is provided by the case of a well-known Australian psychologist, Donald Thomson. He had appeared on a live TV broadcast on eyewitness testimony along with the local chief of police and other dignitaries, and in the program Thomson had described the salient features of identification, using his own face as an example. Several days later he was picked up by the police and charged with rape! This was despite his protestations that he had participated in a television program on the night in question. "Along with the chief of police!" he told his captors, who were unimpressed: "Sure . . . and with Jesus Christ, and who else?" they asked sarcastically.[1] It turned out that the actual perpetrator had broken into a woman's apartment and raped her while she was watching the TV program and had viewed Thomson's face during the traumatic event. She provided the complete details of his description to the police, who then arrested Thomson. He was speedily released, of course, but the incident illustrates how a bystander's features may be mistaken for the perpetrator's features in the course of a violent event, and then registered as a vivid and emotionally charged memory.

Much research has been done to show the importance of emotion for subsequent memory. Some studies have investigated memory in real-life situations, as in the above case, whereas other research has been done in the laboratory. Results from laboratory experiments have shown that emotion-provoking words and pictures are better remembered than are neutral ones. Laboratory experiments are less compelling than the findings from real-life incidents, but they provide greater control, allowing the measurement of attention and other contributing influences on memory. Studies have also explored the neural bases for emotion, including the roles of the amygdala, hippocampus, and orbitofrontal cortex. The amygdalae (one on each side of the brain) are involved in memory for emotional events, whereas the hippocampi are involved in memory for both emotional and neutral events. Interactions between the two areas as well as their interactions with the orbitofrontal cortex are of great current interest. Discussion of these and other neural bases of memory is covered in chapter 7.

Results from such experiments reveal that memory is not indelible but rather depends on the focus of attention along with the engagement of emotion-processing regions during memory encoding. Arousing stimuli can capture and sustain attention, leaving fewer cognitive resources for the processing of other event details. As initially described by J. A. Easterbrook in 1959, arousal narrows the focus of attention such that people remember the central emotional content of an event but often forget other details. Following this lead, a study done by Elizabeth F. Loftus and colleagues presented the participants with a series of slides depicting an event in a fast-food restaurant.[2] Half the participants saw a person point a gun at the cashier, whereas the other half saw a person hand the cashier a check. Eye movements were recorded while the participants viewed the slides. Not surprisingly, the eye fixations on the weapon were found to be more frequent and of longer duration than were fixations on the check. Memory for the slide sequence as a whole was poorer for the participants in the weapon condition than for those in the check condition. For the same reasons, witnesses to a crime often remember the weapon but not other details such as the perpetrator's face or clothing.

Such trade-offs are more likely to occur when there is an object such as a gun that "grabs attention" rather than when emotion is induced only as a general theme. For example, Cara Laney and colleagues found that when participants listened to a story about either date rape or a successful first date, the participants who heard about the date rape showed better memory for all aspects of the story than the individuals who heard about the first date, with no trade-off elicited.[3] This suggests that trade-offs occur only when there is an "attention magnet." In the absence of this type of emotional focal point, the enhancing effects of emotion are more widespread.

Over time, there is an increased advantage for emotional over neutral material. Tali Sharot and Elizabeth A. Phelps found no difference between emotion-provoking and neutral words on an immediate test of recognition memory, but did find a difference in favor of emotional stimuli after a twenty-four-hour delay.[4] Such an increased advantage for emotional material has been attributed to the benefits of enhanced *consolidation* (the gradual fixation of memory traces) into long-term memory. The effects of consolidation on long-term memory are discussed further in chapter 7.

Although both positive and negative arousing experiences are more likely to be remembered than neutral ones, memories of negative events are typically more vivid. Also, negative events are sometimes maintained with greater consistency or less distortion than are positive events, leading some to argue that a person is sadder yet generally wiser when in an unhappy mood![5] As described previously, negative emotion tends to promote the narrowing of attention and thus memory. In contrast, positive emotion tends to promote a broadening of attention and memory along with increased vulnerability to reconstructive memory errors. Negative memories are judged to be more remote than they should be, whereas positive memories are judged as occurring more recently than the actual event. We tend to judge our triumphs as having happened in the near past, whereas our failures are thought of as taking place in the distant past, and sometimes even belonging to a different "me."

As described in chapter 3, memory retrieval depends on the person's present mental "state" (in this case, mood), such that happy memories are more likely to come to mind when one is happy, whereas negative memories are more likely to come to mind when one is sad. In a reversal of this process, emotional memories can serve to influence a person's present mood. Given that the retrieval of emotional

Negative events are sometimes maintained with greater consistency or less distortion than are positive events.

memories can bring back some of the emotions tied to the original experience, it is possible to selectively recall positive memories and so enhance mood.

Remembering Emotion

Remembering the emotion associated with an event can serve as a shortcut, allowing people to decide whether they would want to repeat the experience in the future. For example, "gut feelings" can guide decisions to seek out or avoid a person or situation. Consequently, it is important to determine the accuracy or distortion of those memories over time. One review showed that people remember their emotions fairly accurately, in that the intensity reported initially is highly correlated with that reported after a delay.[6] Such accuracy is not always the case, however.

One predictor of how people will recall past emotions is their current emotional state. For instance, when widowers or widows reported the intensity of their grief following the death of their spouse, they noted much more intense grief after six months than after five years. Yet when asked after five years to state how much grief they had felt six months after their spouse died, the reported level of grief was more highly correlated with their level of grief after five years than with that reported after six months. As in other domains, reconstructive processes are a critical factor when remembering emotions, serving both to enhance memory and to make remembering more vulnerable to errors.

Flashbulb Memories

What were you doing when you learned that Abraham Lincoln was assassinated? The event was unlikely to have been a significant one for you. It obviously *was* a significant event for people at the time, though, and their experiences were reported in 1899 in a study by psychologist F. W. Colegrove.[7] He found that people reported their whereabouts including small details of the occasion with great confidence although thirty-three years had passed. Much later, in 1977, Roger Brown and James Kulik asked people to report what they were doing when they heard that John Kennedy was shot in Dallas.[8] They concluded that there is "hardly a man now alive" who cannot recall the circumstances in which they learned that Kennedy was assassinated. Such detailed reports were termed "flashbulb memories" by the investigators, who commented that it was as if a flashbulb had gone off capturing a fine-grained picture of the details surrounding the event. They proposed that these highly emotional, vivid memories might be caused by a different mechanism from the processes underlying the formation of other autobiographical memories. They labeled the mechanism *print now*, underscoring the arguably indelible, vivid, and elaborated nature of flashbulb memories. They also suggested that such "flashbulbs" are more likely for events that are personally more consequential. In support of this claim, they found that only thirteen of forty Caucasian participants had a flashbulb memory for hearing that Martin Luther King Jr. had been assassinated, whereas thirty of forty African American participants reported that they did have such a memory.

Ulric Neisser cast doubt on the validity of such supposed flashbulb memories by describing one of his own.[9] He wrote, "For many years I have remembered how I heard the news of the Japanese attack on Pearl Harbor, which occurred the day before my thirteenth birthday. I recall sitting in the living room of our house—we only lived in that house for one year, but I remember it well—listening to a baseball game on the radio. The game was interrupted by an announcement of the attack, and I rushed upstairs to tell my mother." He goes on to say that the memory had gone on so long and was so vivid that he never questioned it until he realized its absurdity. In particular, it dawned on him that of course nobody broadcasts baseball games in December, which was the month in which the attack on Pearl Harbor occurred.

More formal evidence suggesting that unlike a photograph, flashbulb memories may often be invalid was provided by the results of a study by Ulric Neisser and Nicole Harsch on recollections of the space shuttle *Challenger* explosion in 1986.[10] The morning after the event, first-year high school students wrote a description of how they heard the news, and then answered questions based on the categories of experience used by Brown and Kulik, such as: What time was it? How did you hear about it? Where were you? What were you doing? Who told you? How did you feel about it? Two and a half years later, the same students, now seniors, wrote a further description of how they had heard the news and answered the same questions about their experience. The results revealed large differences between the original and later reports for many of the participants. A group of judges measured the consistency of the reports by estimating the extent of such changes; they found that the consistency was generally low, with a mean score of 2.95 out of a possible 7. Eleven of the forty participants had a score of 0, having changed their answers to all the questions! Despite the changes, the participants were highly confident in the accuracy of their reports.

Flashbulb memories are defined as those related to *learning about* some shocking event, and a number of studies have compared these memories with "event memories" for the objective details of the occurrence (e.g., that there were four airplanes involved in the 9/11 attack). One such investigation examined flashbulb memories for the 9/11 attack over a ten-year period.[11] Notably, that investigation included fifteen coauthors, allowing for the comparison of reports from different geographic regions. As it happened, large differences across geographic areas were not found except that flashbulb memories were more likely in New York City than elsewhere. The results showed that both flashbulb and event memories declined rapidly across the first year, but did not do so appreciably over the following years. As in other studies, the consistency of reported flashbulb memories declined over a ten-year period, but the participants' confidence in their accuracy remained high, whereas the confidence for details of event memories declined. The inconsistencies that occurred during reports of flashbulb memories were likely to be repeated in later reports rather than corrected. Inaccurate event memories were quite likely to be corrected, however, possibly as a result of people viewing media reports.

Studies have also attempted to pin down the crucial components of flashbulb memories, with likely candidates including surprise, the distinctiveness of the event, the consequences for the person, and their resulting emotional state. In the words of investigators, "Consistent findings have proven elusive," but it seems to us that the major recurring characteristics of such memories are surprise and shock, and often an incident concerning some well-known public figure. Commenting on the nature of flashbulb memories, Neisser wrote that two narratives that are normally kept separate—the course of both history and our lives—are momentarily put into alignment. One widely agreed-on difference between everyday autobiographical memories and flashbulb memories is that confidence in flashbulb memories remains high despite the decline in their consistency, whereas event memories decline in both consistency and confidence over time. The vividness, elaborateness, and ease of retrieval of flashbulb memories likely account in part for the high confidence assigned to flashbulb memories. Again, memory is not indelible although it is sometimes thought to be so.

Involuntary Autobiographical Memories

As described in the preface, memories sometimes involuntarily come to mind, as in novelist Marcel Proust's description of the memories that flooded through him after tasting a little madeleine cake moistened with tea. Involuntary autobiographical memories typically "pop to mind" as opposed to the deliberate retrieval associated with most memories. In clinical psychology, involuntary memories have been given substantial attention, almost always in relation to stressful negative events such as those associated with post-traumatic stress syndrome.

Involuntary memories occur regularly for most people, however, and indeed in 1885, Hermann Ebbinghaus listed involuntary memory along with voluntary and unconscious modes of retrieval as the most common forms of memory. As with voluntary remembering, involuntary remembering favors the recollection of past events with distinctive features that match those of the current context, or past events that are highly accessible due to their novelty and the emotions they provoke.[12] The important difference between voluntary and involuntary memories is the way they are retrieved. Voluntary remembering is a goal-directed process that relies on cognitive control, whereas involuntary remembering involves little or no executive control, but depends largely on associative processes. This lack of executive control results in involuntary retrieval happening faster than voluntary retrieval.

Further evidence for two forms of retrieval comes from a neuroimaging study comparing voluntary and involuntary recall. The results showed that voluntary recall was associated with enhanced activity in the areas of the right prefrontal context concerned with strategic control, whereas involuntary recall was associated with activity in the left prefrontal cortex and other brain regions.[13] Voluntary remembering is thus driven by our intentions, whereas involuntary remembering is associative, driven by details of the current situation. On many occasions, we find ourselves suddenly thinking about some past occasion for no apparent reason. Although the connection to the present context is frequently not obvious, it is most likely that there *is* a match, albeit at a level below our conscious awareness.

Studies have demonstrated that involuntary remembering occurs nearly as often as voluntary remembering and that involuntary memories tend to emerge when attention is unfocused. Although involuntary memories may be maladaptive as in the case of post-traumatic stress syndrome, they are typically helpful and adaptive. They serve to free us from living only in the present, allowing us to learn from the past and imagine possible future events.

The change of focus from associative mechanisms to more cognitive ones such as organization and schemata was discussed in chapter 1, and has been described as the *cognitive revolution*, but it might be better portrayed as the *inclusion* of cognition in a more comprehensive framework. Association as a means of bringing memories to mind plays an important role in guiding behavior and serves as a less strategic basis for retrieval. Indeed, the distinction between recollection and automatic influences of memory noted in chapter 4 corresponds to a description in terms of different bases for retrieval. In the verbal learning tradition, consciousness played little or no part in bringing associations to mind. More recently, much has been written about the "new unconscious" with respect to both social psychology and memory. The authors argue that the new unconscious is responsible for unaware influences of

Involuntary remember-
ing occurs nearly
as often as voluntary
remembering and
involuntary memories
tend to emerge when
attention is unfocused.

memory of the sort mentioned in chapter 4 as well as being involved in memory (mis)attributions. Next, we turn to the "old" unconscious proposed and analyzed by Sigmund Freud, and question whether repressed memories can be recovered by means of psychotherapy.

Memory Wars

Freud proposed that repression is a defense mechanism serving to prevent traumatic memories from becoming conscious, and that the repression of such traumatic memories has serious consequences, including physical symptoms and psychological problems. He therefore argued that the recovery of traumatic memories, making them available to consciousness, would alleviate or even remove the problems they caused when repressed. In the 1990s, there were several widely publicized cases of recovered memories. Typically the cases involved a woman who uncovered a memory during therapy of having been sexually abused during childhood by a family member or authority figure, such as a priest or teacher. Some of the cases involved celebrities including a former Miss America and comedian Roseanne Barr.

A prominent case was that of George Franklin, who was convicted of murdering nine-year-old Susan Nason, a friend of his daughter. The conviction was largely based on his daughter's recovered memory of her father having committed the murder. The case was subsequently overturned on appeal. Unsurprisingly, the affair had a devastating effect on Franklin's family relationships as well as other aspects of his life.

Controversy surrounding such cases arose due to the psychoanalytic techniques used to recover the allegedly repressed memories. The therapeutic techniques included hypnosis and guided imagery along with encouragement to retrieve the repressed memories that the therapist held responsible for the client's problems. A danger created by such techniques is that they can create false memories. For example, research has shown that hypnosis increases the susceptibility to suggestion, increasing the likelihood that both true and false memories will be be brought to mind and described as vivid, but does not provide a means of discriminating between the two.

It seems likely that at least some supposedly recovered memories are false ones. Among the recovered memories are ones of sexual abuses in satanic cults and even alien abductions. Clearly there are good reasons to suspect the validity of such memories. Although thousands of individuals have reported ritual acts of sexual abuse, not a single case has been documented despite extensive investigations by local legal authorities and the FBI. Also, as discussed earlier in this chapter, high emotion typically results in *enhanced* memory rather than in repression.

It seems likely that at least some supposedly recovered memories are false ones.

Concerns related to the possibility of false memory resulted in the creation of the False Memory Syndrome Foundation, backed by research by cognitive psychologists showing that false memory can be produced by therapeutic procedures aimed at recovering repressed memory. Additionally, there have been cases in which people who experienced recovered memories later concluded that the supposedly recovered memory was a false one. Clinicians and others have responded to such criticisms, resulting in the "memory wars."

Experiments have shown that false memories can be created by means of procedures like those employed by psychotherapists to encourage recall of a traumatic experience. In a dramatic demonstration by the US psychologists Elizabeth Loftus and Jacqueline Pickrell, adult participants were given brief descriptions of childhood events and asked to work toward remembering them.[14] Three of the four event depictions given to them were about true events provided by their parents, whereas the fourth was a fictional narrative of having been lost in a shopping mall. After attempting to recall the fictitious event, many of the participants came to believe that the event had actually happened and even produced elaborate details of the incident. For instance, one participant believed that he had been assisted by a kindly old man. Descriptions of these and related studies are included in a review article written by Loftus.[15]

Numerous experiments have since found that false memories can be implanted using similar methods. The incidence of false memories varied across the experiments, probably reflecting differences among the participants, variations in event plausibility, and the amounts of memory reconstruction undertaken by the participants. Some experiments have been highly successful in implanting a false memory, particularly when a photo was presented as a cue for remembering. In one example (a study by Stephen Lindsay), the participants were shown a faked photo of them taking a balloon ride when they were children and asked to recall the event using visually guided imagery. The results revealed that 50 percent of the participants experienced a false memory of the fictional ride. The participants often expressed high confidence in their false memory and great surprise when informed that their memory was false when debriefed. The effects of providing photos are important because therapists commonly encourage clients to look at photos as a means of cueing repressed memories, and such instructions might increase the probability of false remembering.

Cases of recovered memory are typically not corroborated by others, although there are a few reports of recovered memory that turned out to be genuine, including cases in which the abuser implicated in the memory later confessed. One such instance was reported by Jonathan

Schooler and colleagues; it involved a woman who in the course of therapy gradually remembered that she had been raped.[16] Subsequently, she not only remembered having been raped but also remembered having gone to court to help prosecute the person who had attacked her and that he was found guilty. Nonetheless, she had forgotten the incident over several years. She later mentioned the prior rape to her husband, believing that she had never told him about the incident, although as it turned out she had done so earlier.

Are such cases due to repression or the normal mechanisms of forgetting? It is certainly true that when something painful happens to us, we try not to think of it, with the result that the event becomes difficult to retrieve unless the cues and context are relevant. Research has also shown that false memories can be induced by suggestion and then stabilized by the person continuing to think about their contents over time. Confidence in these false memories can be high despite their being created by the mind rather than by external stimulation. As cognitive psychologists, we authors have difficulty fitting repressed and recovered memories into what else we know about cognitive systems, but we are not clinicians, and it is certainly possible that traumatic events affect normal systems in unusual ways. What is needed is more common ground between clinical practitioners and cognitive experimentalists, and more research to disentangle cause and effect in this extremely sensitive area.

Eyewitness Testimony: Malleability of Memory

Eyewitness testimony is generally treated as a compelling basis for conviction in court cases. As described above, however, emotion provoking events can result in false memory that is held with high confidence. Its doing so can have catastrophic consequences. Recent studies show that 3 to 6 percent of all people incarcerated in US prisons were wrongfully convicted. The advent of DNA testing in the late 1980s provided a new level of accuracy for determining the guilt of people incarcerated for a crime and consequently revealed several cases of wrongful conviction. Of those cases, 71 percent resulted from erroneous eyewitness testimony. Among those are cases of a mistaken description of the perpetrator of the sort discussed at the beginning of this chapter as well as wrongful lineup identifications. Of those misidentifications, 41 percent involved cross-racial misidentifications, the vast majority of which were the misidentification of African Americans. In that vein, research has shown that recognition is generally more accurate for persons of the same race and age as oneself. Further, witnesses of a crime or accident are sometimes later misinformed about some aspects of the event as a result of talking with others, including police and lawyers.

The Misinformation Effect

In an experiment done by Elizabeth Loftus, David Miller, and Helen Burns, the participants first viewed a slideshow of a traffic accident with one of the slides depicting a car at a stop sign.[17] After viewing the slideshow, the participants were asked several questions about events in the slides. Among those was a misleading suggestion that the car stopped at a yield sign even though the slides had shown a stop sign. When later tested on the information viewed in the slides, the participants receiving the misleading question were much more likely to mistakenly claim having seen a yield sign than were those not receiving the misleading question, thereby demonstrating a misinformation effect.

Elizabeth Loftus and John Palmer found that the wording of a question could influence the likelihood of people saying they had seen things that did not actually occur.[18] Those participants who were given questions indicating that the cars in a film *smashed* into one another were more likely to report having seen broken glass in the film than were those whose question suggested that the cars *collided*, *bumped*, *contacted*, or *hit*, even though there was no broken glass in the film. Also, the speed of the cars in the film was estimated to be much higher when the question implied that the cars in the film *smashed* into one another.

Many subsequent experiments have shown misinformation effects. The participants have been induced to report seeing a variety of objects that did not actually appear, including a hammer, a mustache, and even a barn. Among theoretical accounts of misinformation effects is one that holds that such effects arise from a misattribution of familiarity of the sort found in the false fame experiments reported in chapter 4 that placed recollection in opposition to familiarity. Stephen Lindsay made use of an opposition procedure to investigate the misinformation effect.[19] In his experiment, the participants saw a slideshow followed by a narrative in which misinformation was included. The participants were correctly informed that there was no question on the test for which the correct answer came from the narrative. Thus the misinformation effect would occur only if the participants failed to recollect that the misinformation was presented in the narrative and hence misattributed the familiarity of the misinformation to the slideshow. The situation parallels that described for the opposition of recollection and familiarity in the false fame experiments noted in chapter 4. The conditions were manipulated in Lindsay's experiment such that in a high discriminability condition, but not in a low discriminability one, it was easy to recollect the source of misleading information. More errors from the misleading source were found in the low discriminability condition. Based on the results from false fame experiments, one would expect to find that misinformation effects would be more likely when people are required to respond quickly or under conditions of divided attention, and more likely for older than for young adults. Such effects have been found, showing the importance of cognitive control for avoiding misinformation effects.

False Confessions

Surprisingly, 28 percent of the convictions overturned by DNA evidence involved false confessions. Why would people confess to having committed a crime that they did not commit? The reasons for doing so can be illustrated by examining the famous case of the "Central Park Five." The five were African American teenagers who falsely confessed to raping a jogger in Central Park in New York City. After prison sentences ranging from six to thirteen years, they were released after a serial rapist whose DNA matched that found on the victim admitted to having committed the crime. The boys' false confessions that resulted in their conviction were obtained during interrogations lasting up to thirty hours. Only the confessions were recorded on video, not the hours of interrogation that led to them. Subsequently, the police admitted that they lied to one of the boys when questioning him that his fingerprints had been found on the shorts of the jogger. Also, the boys were told that they would be given lighter sentences if they confessed. None of the boys admitted raping (that is, penetrating) the victim. Rather, each of their statements minimized their own involvement, placing the blame on others. The boys were convicted, although taken together, the stories told in their confessions made no sense. It seemed as if they were talking about different crimes.

The story of the Central Park Five is as much one of social injustice as of false confession. Nevertheless, it does illustrate conditions relevant to the latter. The young are more prone to false confession. The innocent are less likely to call for the presence of a lawyer than are the guilty, believing that calling for a lawyer would make them appear guilty. Long hours of interrogation with no breaks occur in drab rooms. Promises of lighter sentences are offered to encourage confession. This form of interrogation is a sort that encourages false memory.

Clearly eyewitness testimony cannot be eliminated as a basis for conviction, and confessions of a crime cannot be fully discarded. Yet steps can be taken to increase the certainty that testimony is not contaminated. For example, initial testimony is less likely to be contaminated than testimony given later because of forgetting during the interim as well as the effects of talking with others during the interim. Memory is malleable, thus underlining the importance of gathering converging evidence from several sources.

AGING AND MEMORY

Problems with memory are high on the list of complaints that most people have as they grow older. Are memory losses an inevitable accompaniment of the aging process? Our short answer is that some degree of decline in memory performance is usual and entirely normal in the course of healthy aging; it is *not* a sign of encroaching Alzheimer's disease or other neuropsychological conditions. We will make that distinction clearer toward the end of the chapter. It is also the case that age-related losses are much less in some forms of memory than in others. The purpose of this chapter is to survey what we know at the present time.

Researchers have taken various approaches to understanding these differences in age-related memory difficulties. One obvious change in cognitive processing as we age is that the speed of processing declines from the mid-twenties on. This is shown clearly in the time it takes to react—braking in an emergency, for example—and Timothy Salthouse and his colleagues have shown that deficits in age-related memory performance can be attributed to this general slowing.[1] A second approach was suggested by Lynn Hasher and Rose Zacks, who drew the distinction between automatic and effortful memory processes.[2] They proposed that automatic processes are those that are either genetically "wired in" or have been fully learned; these processes typically run off without intention, require minimal attention, and so do not interfere with other ongoing cognitive activities. Such automatic processes change little in the course of aging. In comparison, effortful processes such as deliberate learning and retrieval operations do demand considerable amounts of attention, interfere with other cognitive activities, and decline markedly with age. More recent work by other investigators has linked such effortful operations to the involvement of cognitive control processes, as discussed later in the present chapter. A third approach has focused on the nature of various memory processes and their control systems—processes such as encoding, retrieval, and working memory—and this is the type of research carried out by the authors of this book.

A dramatic illustration of the different age-related trajectories of various memory operations was provided by Denise Park and her colleagues, and is shown in figure 4.[3] The figure shows the comparative changes in cognitive

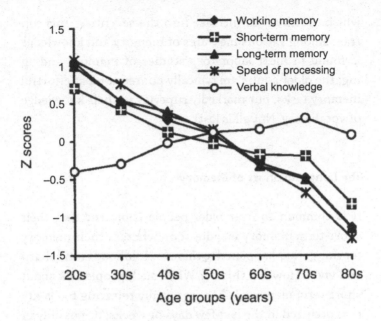

Figure 4 Age-related differences in various measures of memory and cognitive performance, relative to each measure's average performance level, shown in the graph as 0. Reproduced from D. C. Park, G. Lautenschlager, T. Hedden, N. S. Davidson, and A. D. Smith, "Models of Visuospatial and Verbal Memory across the Adult Life Span," *Psychology and Aging* 17 (2002): 299–320, with permission from the American Psychological Association.

processes from the twenties to the eighties; the somewhat shocking point is how early the effortful (or cognitively controlled) processes start to decline. Subtle changes can be detected by the age of thirty. The exception is verbal knowledge (vocabulary knowledge and the use of words), which continues to increase into the seventies. This contrast among various measures of memory and knowledge is found in most laboratory studies of memory and aging; the older adults are typically poorer at most effortful memory tasks, but markedly superior in their knowledge of vocabulary. Not all is lost!

The Long and Short of Memory

It is common to hear older people comment that their short-term memory is quite poor, whereas their memory for things that happened to them in childhood is still sharp and vivid. How can this be? When such people talk about "short-term memory," they are usually referring to events that occurred in the last few days or weeks, or possibly in the last few minutes if they forget what it was they came upstairs to fetch. As we discussed in the first few chapters, however, the distinction that makes more sense from the results of laboratory studies is that between information held in mind and information that must be retrieved to bring it back to conscious awareness. The first category is referred to as "primary" or "working memory," and the second as "secondary" or "long-term memory."

Researchers have found that age differences are minimal in the amount of information that can be held in mind or maintained in working memory, provided that it is not

The shocking point is how early the effortful (or cognitively controlled) processes start to decline. Subtle changes can be detected by the age of thirty.

transformed or otherwise "worked with." So, for example, a person in their seventies can listen to and then reproduce a seven-digit telephone number as well as someone in their twenties. When the limit is tested for the longest string that a person can recall immediately, age differences do emerge, but the difference is small—perhaps eight digits in the twenties and seven digits in the seventies. Interestingly, amnesic patients are also able to maintain as many words or numbers in working memory as healthy control participants. Even the densely amnesic patient H. M. described in chapter 1 could recall short lists of words or numbers, but if asked for the words again after being distracted by other procedures, he would look blank and ask, "What list of words?"

Things are somewhat different if the information in working memory must be manipulated or transformed in some way, and this is the steep age-related decline shown in figure 4. For example, we have used a test called "alpha span" in which the participants are read a short series of words, and the task is to say them back in the correct alphabetic order.[4] So the sequence "table," "cabbage," "iron," "sugar," "lemon" should be recalled "cabbage," "iron," "lemon," "sugar," "table." When this test is used to find the longest string of words that a person can transform mentally and then recall in this way, the number drops reliably from the twenties to the eighties.

The research suggests that age-related memory problems for recent events are attributable to the reduced efficiency of both encoding and retrieval processes in secondary or long-term memory. Once information is dropped from conscious awareness, it must be recalled from long-term memory—and the characteristics of this form of memory are basically the same after an interruption of thirty seconds as after a delay of months or years. At the time an event is experienced, the resulting mental processes are typically less vivid and elaborate in older adulthood; that is, "encoding processes" are less efficient. Retrieval processes are also less efficient, largely due to poorer cognitive control processes reducing the effectiveness of recollection. In turn, this reduced effectiveness is attributable to the reduced efficiency of frontal lobe mechanisms, as discussed in chapter 7. These reductions in efficiency affect the encoding and retrieval of newly experienced events, and also impair the retrieval of events that happened some time ago. As well as affecting the recollection of specific events (episodic memory), aging reduces the ability to retrieve facts and items of general knowledge. Such factual knowledge (including vocabulary) is generally well *retained* by older adults, but is sometimes difficult to *access* (e.g., problems retrieving specific words and names that we know perfectly well). These difficulties can often be resolved by providing more associations, cues, and reminders, and by reinstating the context in which a word or fact was acquired.

Self-Initiated Activities and Environmental Support

At the beginning of this chapter we mentioned that although memory performance typically declines in the course of aging, the age-related lapses are greater in some memory tasks than in others. One typical laboratory task is to read a list of twenty unrelated nouns, one at a time, to the participants to recall in any order. In this "free recall" task, the usual finding is that older and younger adults recall the same number of words from the end of the list—the words they have just heard—but also that older adults recall fewer words from the first fifteen or sixteen words presented. The equivalent recall from the end of the list accords with the finding of minimal age differences in working memory (given that the words did not have to be transformed in any way). The age-related drop in recall from the beginning and middle of the list reflects the difficulty of recollecting information that has dropped out of conscious awareness, despite the point that these words were clearly *in* the participant's awareness only seconds earlier.

If the participants are given a recognition test for the words they failed to recall, their performance levels improve dramatically, especially for older adults. For example, if a person recalled eight out of twenty words in the first test, we take the twelve unrecalled words, mix them randomly with twenty-four new words, and ask the participant to identify words from the original list. The typical finding is that an older adult will now recognize six or eight of the twelve. This laboratory result corresponds to the everyday experience that a cue or reminder can often result in the recovery of a temporarily forgotten name, word, or event. The information was "in memory," but there was a failure of retrieval.

Results like these led Craik to suggest the ideas of *self-initiated activities* and *environmental support*. In line with other researchers, Craik proposed that the processes of remembering, like those of perceiving, depend on both information generated in the brain and information received from the external environment.[5] In the experiment described above, the extra information supplied in the recognition test may have reminded the participant that the word "walrus" was on the list ("I now remember thinking of Californian beaches when I heard the word").

On the other hand, obviously we can recall events when we are in a different context; we can recall a day at the beach last summer even while in the city in January. Craik referred to such memories generated out of context as dependent on self-initiated activities. He further suggested that self-initiated activities and environmental reminders are complementary; less environmental support means that greater numbers of self-initiated activities are required. He also proposed that aging is associated with a decline in the efficiency of self-initiated activities, and that older adults therefore need more environmental support for memory to work at optimal levels.

In a simple demonstration of this principle, an experiment presented thirty-two objects to younger and older adults in a memory task; sixteen of the objects were presented as words and the other sixteen as pictures.[6] The participants were asked to recall as many as possible of all thirty-two items. Following the recall task, a recognition test was given—the original thirty-two items plus thirty-two new items, now all presented as words. Figure 5 shows the percentages of words and pictures recalled and recognized by the two age groups.

The figure depicts that recognition levels are higher than recall levels, pictures are associated with better recall than words, and in general young adults recollect more than older ones. These effects were all expected from previous studies. More of interest, the performance gap between young and old participants declines from left to right on the graph. For example, young adults *recalled* 33 percent of the words, compared to 17 percent for older adults—almost twice as many—whereas older adults *recognized* 83 percent of the pictured items compared to 84 percent for the younger group—almost identical levels of performance.

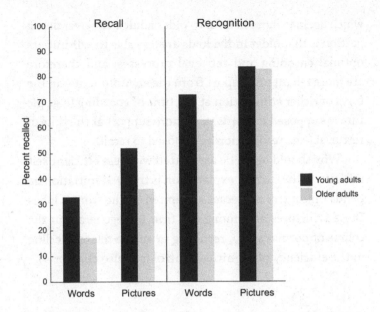

Figure 5 Objects were presented as words (names) or pictures. The figure shows the percentages of objects recalled and recognized by young and older adults. Data from F. I. M. Craik and M. Byrd, "Aging and Cognitive Deficits: The Role of Attentional Resources," in *Aging and Cognitive Processes*, ed. F. I. M. Craik and S. E. Trehub (New York: Plenum Press, 1982), 191–211. Figure by Lindsay Craik.

Our proposed explanation is that pictures provide a richer, more elaborate mental representation than do words, and this is particularly helpful for older people who do not encode elaborately in a spontaneous fashion. Similarly, at the time of retrieval, recognition memory (where test items are re-presented) supplies more environmental support, which again is more helpful to older adults. The overall suggestion is that older individuals are less able to self-initiate optimal encoding and retrieval processes, and therefore are more reliant on support from the environment—in the form of richer stimulation at the time of encoding (e.g., pictures as opposed to words) and more support at the time of retrieval (e.g., recognition as opposed to recall).

Why should aging be associated with less efficient self-initiation? The current explanation is that self-initiation depends on control processes generated by the frontal lobes. These structures are among the first to degenerate in the course of normal aging, resulting in an age-related decline in the efficiency of cognitive control (see also chapter 4).

Forms of Memory, Aging, and Control

If a reduction in the effectiveness of control processes is a major factor in cognitive aging, what effects should we expect to see in the various forms of memory we have discussed in this book? The basic principle ought to be that age-related deficits should be most evident in the types of memory most dependent on control, with minimal effects of aging on tasks and processes in which cognitive control plays a minor role.

We have just described an experiment involving the conscious recollection of recently presented words and pictures.

In general terms this laboratory task is tapping *episodic memory*—memory for events that happen to us personally, ranging from the relatively trivial (learning words in a lab experiment) to the profound (some traumatic emotional happening). The recall of episodes clearly does involve control, as shown by the increase in performance when control is made less necessary by the use of recognition tests.

Another form of memory that involves control in varying degrees is working memory, mentioned earlier in the chapter. Alpha span (where a short list of presented words must be reordered alphabetically) requires a lot of cognitive control, as does "backward digit span" in which a presented string of numbers must be repeated back in reverse order. I give you 586913 (spoken, one digit at a time), and you should respond 319685. Both alpha span and backward digit span require more cognitive control than simple forward span, and both first-named tasks show greater negative effects of aging.

Memory for facts and abstract knowledge is known as *semantic memory*. Unlike episodic memory, the information in semantic memory is typically not associated with the time and place at which the information was acquired. So we know that Moscow is the capital of Russia, 7 x 4 = 28, and dolphins are marine mammals, but we probably have no idea of where and when we learned these facts. Whether episodic and semantic memory are separate systems is a debatable point among memory researchers, but they certainly have different characteristics. As one example, we described an amnesic patient (K. C.) in chapter 1 whose knowledge of facts and procedures was unimpaired, but he had no episodic memory of any occasion he had used the procedures.

Semantic memory is also relatively well maintained in older adults. Vocabulary knowledge is typically superior in older than in younger adults, as is general factual knowledge of the world. Two reservations in this regard are, first, that facts are often difficult to retrieve if they have not been thought about recently, and this need for periodic refreshment of access may be greater in older adults. The second point is that *names* are clearly instances of abstract general knowledge, but are frequently high on the list of older people's memory complaints. The reason for this difficulty may lie in the *specificity* of name information. We either remember a name or fail to remember it; we can't describe it in other terms. Gaining access to highly specific information in memory is often effortful, especially if the information has not been accessed for some time and presumably requires the involvement of cognitive control processes. We therefore argue that the retrieval of well-known and regularly accessed information from semantic memory requires little involvement of control processes, whereas the retrieval of names and other types of specific information does involve cognitive control, and thus does show age-related declines in performance.

Vocabulary knowledge is typically superior in older than in younger adults, as is general factual knowledge of the world.

Lastly, the unconscious, habitual memory processes examined in chapter 4 do not involve conscious control by definition, and by our present argument should show few signs of aging. In fact, the section on measuring automatic and controlled influences in chapter 4 looked at some experiments that found substantial age-related decrements in cognitive control, but no age differences in the unconscious influences on memory. *Procedural memory* is a related form of unconscious learning and memory that we employ in all aspects of life, from turning off light switches to typing, swimming, skating, driving a car, playing a musical instrument, and many other activities involving perception and associated muscle movements. Much anecdotal evidence suggests that once we master a skill, performance levels change little with age—except perhaps at high levels of skilled performance that require constant practice. This conclusion is also backed up by a number of studies, so in general it seems that the application of well-learned procedural skills involves few control processes and shows few signs of age-related declines.[7]

The Ups and Downs of Older Adults' Greater Reliance on Context

The consequences of age-related decline in cognitive control are not restricted to the effects on memory but rather are revealed in a variety of tasks. For example, less efficient hearing, especially in noisy surroundings, is a common complaint of older adults. The most obvious cause is an age-related decline in the effectiveness of the peripheral and central mechanisms of the auditory system, and indeed this is a major component. But a deficit in cognitive control of the sort found in memory for older adults can play a role too, as the following experiment illustrates.[8] Younger and older participants listened to short sentences that finished with a word presented in background noise. In some trials, the sentence beginning (always clearly audible) predicted the final word (in the noise)—for instance, "She put the toys in the BOX"—whereas in other trials the sentence was misleading—say, "She put the toys in the FOX." The task was always to identify the final word, and the participants were informed that the preceding phrase might be predictive or misleading. In the first case, both younger and older participants were almost always correct; the congruent context facilitated the correct hearing. In the second case, however, the misleading incongruent context resulted in significantly more false hearing errors in the older group.

The same younger and older groups also participated in a memory experiment in which the test phase again included valid or misleading cues to the studied words of the sort noted in chapter 4, illustrated by the story of the hapless professor who flew home after driving to a conference.

When the cues were valid, the older group outperformed the young participants, but when the cues were misleading, the older adults made more false memory errors. As in the hearing study, the older participants were more reliant on the available contextual cues, even when they were misleading. A further important finding in this study is that individuals who were most misled by the inappropriate context in the hearing experiment tended to be most affected by the misleading cues in the memory experiment as well. That is, there was a strong correlation between false hearing and false memory, showing the pervasive consequences of diminished cognitive control. For both memory and hearing, older participants mistook what usually happens for what happened on a particular occasion, relying on more automatic influences of memory rather than on a cognitively controlled basis for responding. This is essentially the same phenomenon described earlier in this chapter as the older person's greater reliance on environmental support in memory situations. When the environmental context is misleading, this demonstrates the "dark side of environmental support" as other authors have called it.[9]

One illustration of this "dark side" is the greater vulnerability of older adults to fake news. According to a recent article, "one of the strongest predictors of engagement with fake news is advanced age." Although other factors, such as less familiarity with social media, come into play in this situation, it seems likely that age-related declines in cognitive control and a resulting greater reliance on presented material are major contributors to the problem.

Less Efficient Retrieval Processes in Older Adults

Much research motivated by the levels of processing approach has emphasized differences in encoding. Yet when retrieval processes are understood as a partial repetition of encoding processes, the notion of "depth of retrieval" becomes relevant, emphasizing the role of cognitive control at the time of retrieval. Larry Jacoby and his colleagues have carried out interesting experiments showing that older adults may not process information as fully or efficiently as younger adults during retrieval processing, even when well supported in a recognition test.[10] The general procedure was that the participants studied words to remember in either a shallow (superficial) or deep (meaningful) manner. In a later recognition test, done separately for deep and shallow words, the deeply processed words were recognized more often than shallow words by both age groups. The tests included new words that had not been presented initially as "foils" to be discriminated from the target words. The experimenters then gave a *second* recognition test in which the foils from the first test had to be discriminated from completely new words. In this second test, young adults recognized more foils from the "deep" first test than from the "shallow" first test, showing that they had processed the foils deeply to discriminate them from target items.

In comparison, older adults did not show a recognition benefit for deep over shallow foils in the second test, implying that they had treated the two types of foils equivalently in the first test—perhaps simply checking if the words felt familiar. In more general terms, the study suggests that older adults' recognition processes are not so deep and searching as the processes carried out by young adults, and this inefficiency is another factor in age-related memory deficits.

Associative Learning and Memory for Source

So far in this book we have mainly discussed memory for individual items and events, but how we connect items together—memory for associations—is an equally important topic. In everyday life we must learn associations between names and faces, signs and required actions, and emotional reactions to certain people and places as well as in many other facets of life. There is strong evidence that this ability to associate two items or two events, and remember the relationship, is impaired in older adults. As we will discuss in chapter 7, one of the main functions of the brain structure called the hippocampus is to form associations between events. Hippocampal function does decline in the course of healthy aging, making the age-related associative deficit understandable.

One key class of situations in which associations are involved is the ability to integrate events with their contexts of occurrence. Knowing where and when some dangerous event took place is crucial to avoiding that time and place in the future. We have stressed the significance of context in remembering; if events are not well integrated with the contexts in which they happen, then reexperiencing the context will be less effective in bringing the original event back to mind.

Given that associative memory is poorer in older adults, it should follow that this form of contextual binding is less effective too, and that memory for where and when events have happened should be impaired. This type of memory is sometimes referred to as memory for source—meaning the source of remembered facts and events. Researchers have established that patients with damage to the brain's frontal lobes have problems with source memory. The frontal lobes also decline in effectiveness in older people, providing a further reason to suspect age-related problems in memory for source. This idea was tested in an experiment involving younger and older adults in Winnipeg (home of the absent-minded professor!).[11] The participants were asked thirty relatively obscure facts about Canadian life; if they did not know the answer to any fact, the answer was provided by the experimenter. One week later, the same participants were given a further test of Canadian trivia; the original thirty questions were mixed with thirty new, somewhat easier questions. For all the sixty questions that they could answer on this second test, the participants were requested to write down where they had learned the fact originally; the options included from TV, radio, a newspaper, friend, school or college course, or "this experiment, last week."

The results showed, first, that older adults knew the answers to more of the thirty new questions presented in week two than did the younger adults—46 to 33 percent. So the older participants were more knowledgeable about the general topic. For the questions that individuals did *not* know in week one (but had been given the answers), both age groups recalled about 40 percent of the correct answers on the second occasion. So memory for new facts was equivalent for the two age groups. Yet for the source memory questions for these newly acquired facts ("Where did you learn it?"), younger adults correctly attributed 89 percent of their answers to the experiment, whereas for older adults the corresponding value was only 56 percent, showing a significant age deficit in memory for the source.

Prospective Memory

The term "prospective memory" refers to forming an intention to perform some action at a future time, and then remembering to carry it out at the appropriate time and place. The concept covers a wide set of tasks, ranging from household trivia (remembering to check the milk in the fridge or put socks in the laundry) to more serious planned actions (remembering to turn off the stove or take medication), and even potentially catastrophic memory failures such as failing to check the dials on a passenger jet or nuclear reactor. The successful fulfillment of a prospective memory task thus involves both remembering that *something* should be done at the planned time or place (remembering to remember), and remembering the actual task to be performed.

What are the factors that contribute to success and failure in prospective memory tasks? US psychologists Gil Einstein and Mark McDaniel have proposed that two main processes are involved in successful prospective remembering: strategic monitoring and spontaneous retrieval.[12] The first is a conscious act in which the person designates a target (a mailbox perhaps or specific time of day) and then monitors the environment until the target is encountered. The second process is more like an unconscious habit, relying on the sight of the mailbox or clock to remind the person of the intention. The researchers have added the notion that these processes can be disengaged when the target could not occur; for example, if the intention is to buy milk on the way home from the office, the relevant processes need not be engaged until the person starts the journey home.

The same researchers have also made the distinction between time-based and event-based prospective memory targets. An instance of the first type of intention would be "I must phone my husband at 4:00 o'clock"; event-based intentions involve a place, person, or occurrence, such as "When you see James, remember to ask him for his wife's telephone number." In general, event-based tasks are more likely to be successful as they involve more tangible reminders, although if the event is one that is commonly experienced (if you meet James every day), it will be less effective. For this reason the best reminders are unusual objects or events, or objects placed in unusual settings ("Now why did I put this book on the hall table? Oh right . . ."). Other researchers have usefully broken down time-based prospective memory into intentions involving a specific time of day (e.g., "Today I have to pick up my son from school at 4:00 p.m."), and those involving a time interval (e.g., "I should take the bread out of the oven in twenty minutes").

Prospective memory is one type of memory that declines in the course of healthy aging, at least partially because many prospective memory tasks involve self-initiation and cognitive control.[13] Age-related deficits are greater in time-based than in event-based tasks for that reason; events provide a reminder, whereas time-based tasks require more self-initiation. Older adults make fewer errors on time-of-day tasks (e.g., remembering to take medication at 6:00 p.m.) than on time-interval tasks, possibly for the same reason. Many older people are well aware of such prospective memory problems and sensibly compensate by making lists, leaving notes, and setting alarms.

One interesting finding in this research literature is that when prospective memory tasks are set in the person's own daily life (e.g., remembering to phone the lab at 10:30 a.m. and 5:45 p.m. each day for a week), older adults consistently outperform their younger counterparts, despite the opposite result (sometimes with the same participants) being typically found in laboratory settings. The reasons for this "age–prospective memory paradox" are still unclear, but one explanation is that older adults may simply be more motivated to succeed in the study. In a sense they may be "batting for the old folks' team," whereas for younger adults the study may be "just another boring psych experiment!" If older adults are more motivated, why do they not also outperform younger adults in lab experiments? The answer may be that in lab experiments, *all* the participants (both young and old) are typically highly

Prospective memory is one type of memory that declines in the course of healthy aging, at least partially because many prospective memory tasks involve self-initiation and cognitive control.

motivated to perform the tasks, whereas in day-to-day life young adults have many other commitments while older adults make the experiment a focus of their day.

Apart from leaving notes and alarms, are there ways to improve prospective memory success? Rather obviously, success rates go up as tasks are more important and failure will have more severe consequences. But in addition, one of the most effective methods is to make use of "implementation intentions." In this procedure, the person simply forms an intention such as "When X happens I will do Y" and may even repeat the intention out loud. For example, if they are planning to receive a COVID-19 vaccination, they might plan doing so in detail: thinking of what they will be doing prior to leaving for the vaccination, their means of transportation, and so on. This simple procedure is surprisingly effective. Indeed, even frontal lobe patients who show a major deficit in cognitive control benefit from employing the procedure.

Finally, why are prospective memory tasks so vulnerable to failure? Most memory tasks are cued by a question such as "What did you eat for breakfast yesterday?" or an instruction like "Please recall the words on the list you studied." Recognition tasks (like identifying a suspect in a police lineup) are even more explicit, in that you are *asked* to remember. But with prospective memory tasks, no such instructions are given at the time of retrieval; the person must simply "remember to remember"—a process requiring self-initiation and cognitive control.

Memory Problems in Healthy Aging and Dementia

When adults in their fifties and sixties start to experience memory problems that they did not have at a younger age, they often worry that this may be a sign of encroaching Alzheimer's disease or some other variety of dementia. It is true that problems with memory are among the first signs of Alzheimer's disease, but as we have shown in the present chapter, an increase in the number of memory failures is a typical aspect of the normal aging process. In chapter 7, we will describe how certain areas of the brain—particularly regions of the frontal and temporal lobes—are associated with memory functions, and it is the case that these areas are among the first to deteriorate in the course of aging. Actually, some regions of the frontal lobes show declines of function by the late twenties; associated cognitive difficulties can be picked up by sensitive behavioral tests (see figure 4). These age-related changes mostly involve the speed with which a person can make complex decisions, but some measures of working memory, such as alpha span (mentally rearranging a short series of presented words to say them back in alphabetic order) also starts to decline slowly by age thirty. Memory for events (episodic memory) is sensitive to aging too, but problems with this type of memory as well as prospective memory typically do not arise until age sixty-five and beyond.

The differences between the mild memory problems associated with healthy aging and those associated with dementia are largely matters of degree. That is, people with dementia experience the same types of problem, but more severely and frequently. So losing your glasses or car keys, forgetting what you came upstairs to collect, and word-finding problems are typical memory failures associated with healthy aging. Such failures can of course happen at *any* age, yet they tend to increase in frequency in the sixties and seventies. In comparison, patients with dementia exhibit such failures as repeatedly asking the same question, forgetting the names of close relatives, getting lost in familiar places, and having problems with activities requiring a sequence of coordinated events and decisions—as in cooking, driving, and using electronic and other household appliances. If these signs occur frequently in a friend or family member, that is the time to get medical advice.

MEMORY AND THE BRAIN

How is memory represented in the brain? It may be worth emphasizing at the outset that memory really *is* in the brain, or at least in the central nervous system. People sometimes talk about "memory in the fingers" with regard to some highly practiced skill such as knitting or playing the violin; in such cases, a skilled activity can run off with minimal attention and cognitive control. Yet we attribute these more or less automatic activities to well-learned habitual representations in the brain linking perception to action. The obvious benefit is that cognitive processes are freed up for other activities; the negative aspect is that if circumstances change (e.g., someone from the United States driving a car on the left-hand side of the road in the United Kingdom), the habitual sequence may lead to inappropriate actions. So deliberate conscious control must be reinstated to monitor behavioral actions and reactions until new as well as more appropriate habit patterns can be established.

Memory in the Brain: Localized or Distributed?

Over the past seventy years or so, research on the neural correlates of memory has gone from a stress on location in the brain to an emphasis on networks. Earlier studies assumed that memories were stored as specific clusters of neurons, whereas more recent research has taken the view that memories are represented by the activity in networks of neurons, spreading diffusely through various brain regions. Recent research has shown that the so-called default network is one deeply involved in remembering the past and planning for the future. It is a network of neurons composed of brain regions in the medial (as opposed to surface) prefrontal cortex, and medial and lateral areas of the parietal lobe (figure 6). The network is most active when the brain is at wakeful rest, and the person is attending to internal thoughts such as daydreaming, mind wandering, and remembering. Given that such musings reflect a person's innermost feelings and previous experiences, it has been suggested that this network contributes substantially to the concept of self—a person's conscious understanding of who they are. It is also the case, however, that different brain regions have been shown to play crucial roles in the encoding (acquisition) and retrieval of qualitatively different facts and events. So perception and memory for visual information (faces and scenes, for example) is located primarily in the occipital lobes at the back of the brain (figure 6).

In the 1950s, US psychologist Karl Lashley argued for a distributed view of memory representations.[1] His experiments had revealed that impairments of learning in animals reflected the *amount* of brain tissue lost rather than the specific location of brain damage. This point of view, though, was questioned by the observations made by the Canadian neurosurgeon Wilder Penfield. In his operations to relieve epilepsy, Penfield would electrically stimulate the brains of conscious patients to find centers of epileptic activity. On some occasions, the stimulation of the exposed areas of brain tissue in the temporal lobe caused the patient to experience vivid sequences of visual and auditory events.[2] The sequences often appeared to include fragments of happenings from the patient's past life, and thus Penfield proposed that autobiographical memories were stored in specific brain areas and were evoked by electric stimulation. Some regions of the brain functioned as a video recorder in Penfield's view.

Later work has thrown this conclusion into doubt, however. The sequences experienced by the patients tended to be general events that might occur in particular settings as opposed to true repetitions of actual lived events. In

Research on the neural correlates of memory has gone from a stress on location in the brain to an emphasis on networks.

addition, the patients described their experiences as having a dreamlike quality, so the current thinking is that these interesting observations are akin to dreams instead of constituting a rerun of actual life events. So the debate about how and where memories are stored in the brain continues into the twenty-first century.

Roles of the Hippocampus

Findings from the study of patient H. M. were pivotal in understanding how new autobiographical memories are encoded in the brain. In chapter 1, we described how this patient had both hippocampi removed surgically (one in each temporal lobe) to deal with his intractable epilepsy. As a consequence, H. M. was unable to form any new episodic (autobiographical) memories for the rest of his life, although he could repeat back a short series of words or numbers, and was able to learn simple procedures. Yet he had no later recollection of learning these skills. These last points make it clear that human memory and learning must involve more than one mechanism, and we discuss possible candidates for these mechanisms later in the chapter. H. M.'s operation was in 1953, and he did learn some general facts about subsequent events; he could answer questions about astronauts, for example, and the

Beatles and President Kennedy, although unreliably and only in general terms.

The hippocampus is not a memory store in itself, though. The current thinking is that various features (colors, shapes, sounds, and movements) of experienced objects, scenes, and events are bound together in the hippocampus to form coherent representations. The integrated representations are then transferred to be stored in relevant areas of the cerebral cortex—the brain's outer layers of neural tissue (the cortex in figure 6). This complex

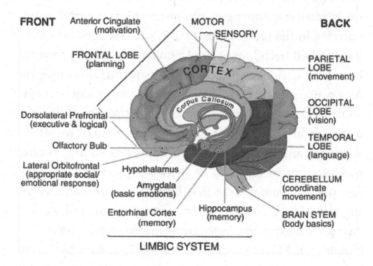

Figure 6 Some of the main areas of the brain involved in memory. Reproduced with permission from Allen D. Bragdon and David Gamon, *Building Mental Muscle* (South Yarmouth, MA: Allen D. Bragdon Publishers, 2016).

series of processes is referred to as *consolidation* of the memory record—a term capturing the idea that the purpose is to make memories stable and resistant to interference. In a first phase, neural circuits in the hippocampus are modified by the growth of new connections and the restructuring of existing connections. A later, slower phase involves a gradual reorganization of the relevant brain areas.

It seems likely that these representations are encoded in the cortex in the same regions that are activated by further incoming perceptions of the world—visual features in visual areas of the cortex, auditory features in auditory areas, and so on. The original "event" is then reconstructed from its constituent components at the time the memory is retrieved. Our suggestion here is therefore that perceiving and remembering utilize many of the same brain structures and processes to give rise to conscious experience.

The hippocampus and associated structures also act to organize "cognitive maps," enabling animals and humans to establish their present location, and find their way to some further desired location on the basis of previous experiences. Such maps are "cognitive" ("smart maps" as it were) in the sense that they function not only as static indicators of present location but as predictive devices too, conveying details of what may be encountered if a specific direction is followed. Recent work with rats has shown that the hippocampus contains "place cells" that fire when the animal is in a specific location. These cells fire in response to odors, touch, and other environmental stimuli as well, so the cells express the rat's location in combination with information about events that took place there. Place cells work together with "grid cells" in the neighboring entorhinal cortex (figure 6); grid cells generate virtual maps of the current environment—a sort of in-built GPS system for the brain.[3]

London taxi drivers are required to undergo extensive training to learn the quickest route between any two places in that large city. This is known colloquially as "the knowledge," and is likely under threat nowadays from GPS devices and other recent technological advances. In 1998, a celebrated study showed that trained taxi drivers had larger hippocampi than normal, and that the size increased with time on the job.[4] It therefore seems likely that humans also possess hippocampal mechanisms that encode spatial and other types of information, again enabling the formation of memory for events that take place in specific environments (i.e., episodic memory).

Hyperfamiliarity: The Case of V. L.

At various points in the book, we have discussed the differences between remembering as a feeling of general *familiarity*, and remembering as a full-blown *recollection* of an event in terms of such contextual details as where and when it occurred, who was present, and other details (see chapter 4). For some time, researchers have asked whether familiarity is simply a weaker form of remembering than recollection, such as a memory experienced with less confidence perhaps. Recent work on the neural correlates of remembering, however, has provided strong evidence that the two experiential states are indeed different forms of memory, mediated by different brain structures and processes.

This area of research is still a work in progress, but the understanding at present is that recollection is associated with activity in the hippocampus along with regions of the frontal and parietal lobes (figure 6). The experience of familiarity is mediated by activity in more lateral regions of the frontal cortex, and also in the perirhinal cortex—a structure near the hippocampus within the temporal lobes of the brain (figure 6). Researchers have examined one patient (N. B.) whose left perirhinal cortex was removed surgically to relieve epilepsy, but whose hippocampus was left intact. The patient lost feelings of familiarity for remembered events, yet was still able to recollect contextual details normally.[5]

One of the present authors (Craik) studied a patient in her eighties known as V. L., who exhibited the opposite set of feelings and behaviors; to her, many events felt extremely familiar although she was experiencing them for the first time. So, for example, she would switch off the TV news saying that she had "heard it all before" (maybe she had of course!). On a car trip with her daughter to an unfamiliar town, they saw a somewhat disreputable man standing on a street corner. V. L. remarked, "That man is always standing on the same corner; you would think that he would have better things to do!" On another occasion while walking on the sidewalk, she was accidentally jostled by a passing jogger; V. L. commented, "She's *always* doing that, she should be more careful!" It seemed as if when unusual objects or events caught her attention, she then experienced them as being highly familiar.

One test we gave her in the lab consisted of a long series of pictures taken from magazines. Some of the pictures were repeated after twenty intervening pictures; the test was to say whether each successive picture was "new" (seen for the first time in the series) or "old" (a repeated picture). We also tested a group of older adults of V. L.'s age and background; they correctly recognized 97 percent of the repeated pictures as old, and on average wrongly identified only 1 percent of the new pictures as repeated. V. L. correctly identified 100 percent of the old pictures as repeats, but also *wrongly* identified 79 percent of the new pictures as repeats. Later in the series, she claimed to have seen almost every new picture before in the series. V. L. was given a structural magnetic resonance imaging (MRI) scan too, and the results showed a loss of brain tissue through atrophy in her frontal lobes as well as in the hippocampus and associated structures, probably including the perirhinal cortex. The study is only on one person obviously, but along with similar cases starts to show how aspects of memory experiences are associated with specific structures and processes in the brain.[6]

Role of the Frontal Cortex

For many years the functions of the brain's frontal lobes remained obscure; patients who had sustained damage to that part of the brain often appeared to live normal day-to-day lives. It has become clearer in the past hundred years or so, however, that many higher cognitive functions (e.g., thinking, problem-solving, and decision-making) are controlled by processes located in different regions of the frontal lobes, and that some of these control functions are associated with memory-encoding and retrieval processes. Understanding the relations between cognitive processes, on the one hand, and brain structures and processes, on the other hand, was greatly accelerated by the advent of neuroimaging techniques—first by positron emission tomography (PET) and MRI, and then by functional MRI (fMRI), which tracks neural changes as they are happening, not just as a static snapshot. These brain-scanning techniques essentially measure changes in blood supply to specific regions of the brain; the logic is that when a particular brain structure or process is activated to carry out a certain cognitive task, the increased neural activity draws blood to that region to facilitate its functioning. These techniques have revolutionized the ability of clinicians and researchers to associate specific brain regions, and damage to them, with intact and impaired cognitive functions.

In a group of studies utilizing PET scanning published in 1994, the researchers investigated the brain regions activated during the encoding and retrieval of verbal information.[7] In order to study encoding, the participants were shown a series of single words (common concrete nouns) while lying in the scanner. In one condition, they decided whether each word contained the letter *a* and responded by pressing one of two buttons. In a second condition, they decided whether each word represented a living thing (e.g., "CAMEL"—yes; "CARPET"—no), and again conveyed their decision by pressing one of two buttons.

The comparison between living/nonliving decisions, which involve semantic processing, and "a-checking" decisions, which involve simple visual processing, is an example of deep versus shallow processing, as described in chapter 2. Deep processing is typically associated with better memory for the material, and this expected result was again obtained in the present experiment. Twenty minutes after the last scan, the participants were given a recognition test for the words they had processed (under both conditions) mixed with new words. They correctly identified 75 percent of the living words but only 57 percent of the a-checking words.

The brain correlates of this memory benefit to deeply processed words was found by means of a computer program that essentially subtracted the pattern of neural activity across the brain associated with shallow decisions from the pattern of activations associated with deep (semantic) decisions. The principal result of this subtraction technique was an activated region of the left prefrontal cortex. So the extra benefit to episodic memory of processing words deeply as opposed to processing them in a shallow manner is associated with brain activity in the left prefrontal region (figure 6).

Another study by the same investigators focused on brain regions associated with retrieval processes in episodic memory.[8] In this study, the participants' brains were scanned while they listened to sentences. Some sentences were novel, and some were repetitions of sentences that the participants had learned the previous day; the task was to indicate which sentences they had heard before. This was a relatively easy task, and successful recognition was again associated with neural activation in the prefrontal cortex, but this time in regions in the lower right-hand side of the brain. So *encoding* for verbal episodic memory is associated with increased blood flow (and presumably increased neural activity) in the *left* prefrontal cortex, whereas episodic *retrieval* of similar material is associated with increased activity in the *right* prefrontal cortex.

One puzzle here is that there is also evidence (discussed previously in this book and later in this chapter) that memory retrieval processes largely recapitulate the same patterns of neural activity that accompanied encoding of the material. Obviously this principle does not fit well with findings of different regions of the prefrontal cortex being associated with encoding and retrieval. The answer to the puzzle is probably that the prefrontal activities are associated with different *control processes* managing the functions of acquisition and retrieval. That is, the actual representations of the material in memory may be located in more posterior brain regions, but the processes associated with the encoding and retrieval of memory are managed and guided by control processes located on different sides of the prefrontal cortex.

One other result is of interest in this connection. A research group at Washington University in Saint Louis have studied a language generation task while the participants are in a PET scanner.[9] In this task, the participants hear a noun and have to generate a meaningfully related verb (e.g., hear "food" and say "eat," or hear "ladder" and say "climb"). In a second task (noun repetition), the participants simply have to repeat the presented noun (e.g., hear "paper" and say "paper"). When the pattern of brain activations associated with the noun repetition task are subtracted from the activations associated with the verb generation task, the results again show a marked region of neural activation on the lower-left surface of the prefrontal cortex. The Washington University group did not test memory for the nouns in its study; the researchers' conclusion was that the left prefrontal activation represented the retrieval of meaning from a verbal knowledge base ("semantic memory"). A group from Toronto repeated the experiment, though, adding a recognition memory test for the nouns, and found better recognition of nouns in the verb generation task (50 percent) than those in the noun repetition task (26 percent).[10]

One way of tying these results together is that the participants must search their general memory store of meanings to generate a verb appropriate to the current noun. It has thus been suggested that the activation in the left prefrontal cortex represents both the retrieval of meaningful associations from semantic memory and the encoding of the same information into episodic memory. That is, the retrieval of meaningful information is a good way to encode it for a subsequent memory test. We described this way of remembering the name of a new acquaintance in chapter 3; the retrieval of a person's name a few seconds after hearing it is a good way to remember the name later.

Memory Retrieval Processes in the Brain

It is a striking fact of human cognition that a specific sight, sound, smell, or verbal utterance can instantly re-create a previous experience in our conscious mind—often an experience we have not thought about for years. How does memory retrieval work? We mentioned earlier and in chapter 2 the idea that the cognitive processes of retrieval essentially recapitulate the same complex set of mental processes that occurred when the event was originally perceived and encoded into memory. If this is so, it should mean that the *neural* processes in the brain associated with retrieval should also largely repeat the processes that took place during the encoding of specific events. Additionally, in chapter 3 we emphasized the importance of reinstating the original context of an event's occurrence as far as possible (either physically or mentally) to remember the event successfully. In general, it does appear from personal experience and many experiments that remembering is successful to the extent that the present set of cues, reminders, and aspects of "environmental support" (see chapter 6) recapitulate features of the original event. We should therefore expect that the neural processes associated with the relevant context should again be activated during retrieval—either driven by the reinstatement of the context itself or filled in by processes of "pattern completion" (see below).

This general idea has received good support from a number of brain-scanning studies using fMRI and other methods. These studies have measured the pattern of neural activation associated with the learning (encoding) of specific types of stimuli—for instance, words presented either visually or auditorily, faces, pictures of objects, musical phrases, and so on—and then in a subsequent phase of the experiment observing the pattern of activation associated with the successful recognition of the original stimuli. As an example, the participants may be asked to remember a series of photographed scenes shown to them while in the MRI scanner. In a second phase, the same scenes are shown again mixed with completely new ones, and the task is to respond "old" by pressing one of two buttons when a previously learned scene is viewed. The result is that the pattern of activated areas of the cerebral cortex associated with successful recognition is largely identical to the pattern associated with the original learning. It should be emphasized that these patterns of activation are specific to the items learned and later recognized. That is, pictures of scenes, faces, and different objects all give rise to different patterns of activation, and each specific pattern is substantially reinstated during successful recognition.

This account leaves out one important feature, namely that in real-life remembering, it is unusual for the reminding context to reinstate the original situation completely.

The pattern of activated areas of the cerebral cortex associated with successful recognition is largely identical to the pattern associated with the original learning.

For instance, the question, "Do you remember that excellent dinner we had in Paris last summer?" does not re-present details of the café, waiters, and general ambiance; somehow the brain re-creates these details, in part at least. To answer this riddle, researchers have proposed that specific brain structures perform operations of *pattern completion*. The idea is that when parts of an original pattern of neural activation is evoked by a retrieval cue (a face, scene, or question), and this partial pattern in turn evokes a feeling of familiarity, the memory system "fills in the blanks" by the processes of pattern completion. Such processes are known to occur elsewhere in perception and cognition. In vision, for example, the retina contains no light-sensitive rods and cones in the region where the optic nerve leaves the back of the eye, yet we do not perceive any gap in the visual field in that location. This is because neighboring receptors complete the expected pattern that is then transmitted to higher visual centers.

One account of how all of this may work draws on research using electroencephalography (EEG).[11] This technique measures the electric currents generated by neural activity and so provides information about the time course of mental events. The researchers propose that the sensory information evoked by a retrieval cue (e.g., a pictured scene or spoken word) is transferred to the hippocampus within half a second. At the same time, other structures near the hippocampus as well as in the frontal cortex generate signals that are consciously perceived as "familiarity" of the word or scene. At this point no further processing may take place, and the person will recognize the object on the basis of familiarity alone. Alternatively, if the person has set the goal of recognizing the object more completely, processes in the hippocampus engage in pattern completion transactions with relevant areas of the cortex for the next second or so to evoke the experience of conscious recollection, in which details of the object's original context of occurrence are reinstated in the person's awareness.

Working Memory

In chapter 2, we described the notion of working memory—the information held in conscious awareness at any given moment. In that earlier chapter, we presented the view that working memory consists essentially of information we are attending to, and that the limited capacity of working memory reflects our limited amount of attentional resources. Although it may feel that the various words, numbers, images, and planned actions that we hold in working memory are all brought to one central place in our minds, the emerging evidence from studies of the brain is that the words, numbers, images, and actions are all represented in their own respective regions in the cortex, and are activated by the processes of attention. So attention "goes to them" as it were, rather than the various different types of representation all converging on a single brain location. The evidence for this conclusion comes from studies using brain-scanning techniques such as fMRI. These experiments have shown that the same cortical areas activated during the perception of words or pictured scenes are again activated when that type of material is held in working memory.

One other brain area involved in working memory operations is the *dorsolateral prefrontal cortex*; this term refers to the upper, side surfaces of regions at the front of the brain (figure 6). This area is associated with cognitive control processes, and working memory is one of the functions controlled. So we can say that when we intend to hold a telephone number or specific pattern in working memory, processes originating in the prefrontal cortex direct attentional resources to the appropriate locations in the posterior cortical regions—the same regions involved in the perception of the material when it was first presented. So in some cases, working memory may be thought of as prolonging perception, but in other cases the information held in working memory is retrieved from more permanent storage sites. For example, if I ask you to hold the product of seven multiplied by eight in mind, or the names of the capitals of France, Germany, and Italy, the required information is never presented explicitly but instead must be retrieved from long-term storage.

Perception, Encoding, and Retrieval: Aspects of the Same Mechanism?

It is important to note that when some stimulus (a word, sentence, picture, or complex event) is initially experienced or deliberately learned, the route and timing of the processing is extremely similar to that previously described for the process of retrieval. The suggestion is that when we perceive an object visually, or a complex sound auditorily, its physical properties are interpreted by the relevant sensory systems in vision and hearing, and "encoded" into the nervous system in terms of neural impulses, which are then transmitted to higher cognitive centers in the brain. These centers first interpret the neural patterns encoded by the sense organs in terms of primitive perceptual features (e.g., lines, edges, colors, shapes, and textures, in the case of visual inputs; pitch, loudness, and timbre, in the case of auditory inputs), and these features are then "bound" together in the hippocampus in consultation with stored representations in higher cortical centers to form representations of known objects, recognized faces, understood words, and so on. So the analysis of this incoming flow of sensory data proceeds in a hierarchical fashion from lower *sensory* levels, through levels of feature extraction, to levels of *perceptual* recognition and meaning, and from there to the highest cognitive levels of *conceptual* understanding and the planning of relevant actions.

In chapter 2, we made the point that the encoding processes in the levels-of-processing framework are simply those of normal perception and understanding. By that account, every perceptual act modifies the cognitive system in miniscule ways, and these subtle changes can later participate in a reactivation of the original perceptual event, provided that the changes are sufficiently distinctive. In the great majority of cases, the changes are *not* sufficiently different from many other encoded changes in the perception/memory system, however, and so like grains of sand on a beach, the record of the original event is "there" yet not retrievable as a conscious memory essentially because it is impossible to differentiate from many other similar records. As we wrote earlier, for an event to be memorable and potentially retrievable, it must be elaborated richly, making it discriminable from other similar perceptual traces.

These are difficult issues to describe and comprehend, but the basic point here is that memory-encoding processes, recording an event in the brain to serve as a memory at a later time, are no different from those involved in perceiving and understanding events in the world around us. The suggestion is that there are no brain processes devoted *exclusively* to memory formation and retrieval; rather, these memory processes utilize many of the same processes used for the perception and understanding of space, the environment, objects, and their interrelations and significance.

In the previous section, we looked at the processes that may underlie recognition memory, when the picture of a face, object, or scene, or an actual person in a police lineup, is re-presented to a person with the question, "Did you see this item before in a specific context?" The context might be a crime scene that the person witnessed or a series of pictures shown to an experimental participant on the previous day. We suggested that in such cases of recognition memory, the presented item is then perceived and processed like any other object in the environment, but in addition, processes in the hippocampus and frontal lobes attempt to integrate the processed event with records of the context in which the event originally appeared. If a sufficient number of previously encoded context records can be successfully integrated with the presented item, this generates the experience of recollection, and the person, object, or pictured scene is recognized as having been seen before in the relevant context.

We need a different depiction of retrieval when considering recall, however. In the case of recognition memory just described, an item to be recognized is presented and then processed by the participant. In the case of recall, the information presented is often a question such as "Can you recall the names of the people attending the office meeting yesterday?" or "What happened after the boy fell out of the boat in the lake last week?" In these cases of recall, rather than processes starting with sensory analysis, and proceeding through feature analysis and object perception to a conceptual analysis of meaning, the processing must *start* at the conceptual level, and then work *down* through the processing hierarchy to reconstruct perceptual and sensory details. So the person recalling the office meeting must first reconstruct details of the location, setting, seating arrangement, and so on, before attempting to reconstruct the details of the people involved. If the questioner went on to ask, "What was Betty wearing at the meeting?" further perceptual and sensory details must be reconstructed. It has therefore been proposed that the processes of memory recall are like those of encoding (and normal perception), but in reverse, starting at a high conceptual level of analysis, and working down through the same processing hierarchy to levels concerned with perceptual and sensory details.[12]

In overview, we are suggesting (along with many current memory researchers) that the processes of perception, memory encoding, and memory retrieval all involve a common set of brain structures and processes that together give rise to the conscious experiences of perceiving, learning, and remembering. The main structures involved in these processes include the hippocampus for binding and integrating aspects of events, regions in the frontal lobes that act to monitor, manage, and control other on-going processes, and the cerebral cortex (the outer layers of the brain), which is the likely site of storage for representations of previous events and the concepts that have evolved from their occurrence.

MEMORY ABILITIES
Excellence, Maintenance, and Repair

The aim of this final chapter is to bring together points from previous chapters to see how they provide a clearer understanding of the factors underlying individual differences in memory performance. Once memory is better understood in this way, it should be easier to develop techniques to enhance it—or at least postpone age-related declines. We begin by looking at cases of extraordinary memory and then describe techniques for enhancing memory with a focus on older adults' memory performance. When discussing cases of excellent memory, we distinguish cases that have a biological basis (from genetic or other sources) from those that reflect cognitive strategies, learned or adopted by individuals. Similarly, memory deficits can reflect biological declines, a change in cognitive strategies, or some combination of the two. As noted later, attempts to repair the memory performance of older adults have taken account of these differences when devising means of rehabilitation. Research findings on the neural underpinnings of memory, as explored in chapter 7, are also providing promising leads for the development of successful rehabilitation techniques.

Cases of Extraordinary Memory: The Mind of a Mnemonist

A number of cases of extraordinary memory have been described over the years, and we will discuss two here. One is a man, referred to as "S" (his name was later revealed to be Solomon Shereshevsky), who was studied by the eminent Russian neuropsychologist Alexander Luria.[1] S was a newspaper reporter in Moscow in the 1920s. His boss observed that S never took notes at meetings or when reporting on speeches, yet he recalled every detail perfectly. When this matter was raised with him, S was apparently astonished that other people could *not* do as he did! Luria tested him with long strings of words, numbers, mathematical formulas, and even poems in languages he did not understand: S remembered them all! He could even recall this information years later; his memory had no apparent limit.

The other salient discovery that Luria made was that S exhibited synesthesia to an extraordinary degree. Information in one sense immediately evoked sensory images in one or more other senses. As an example, S portrayed a speaker as having "a crumbly yellow voice"; he also disliked hearing music in a restaurant, as the music "changed the taste of everything." His visual imagery was *eidetic*, implying that meaning was irrelevant; he could image and remember a meaningless formula just as easily as a page of meaningful text. Unlike almost everyone else, his memory really was "photographic." As we will examine, this irrelevance of meaningfulness is entirely at odds with how most people remember, so his case, while fascinating, is of limited usefulness in our attempts to understand normal memory processes.

Sadly, S's "gift" did not make him happy; on the contrary, he found that his inability to forget anything was disruptive to his life. He was persuaded to become a professional stage mnemonist and made a reasonable living that way. He died in his early seventies from complications related to alcoholism.

A Case of Unusual Autobiographical Memory

Another person with extraordinary powers of memory—but quite different ones—was studied by a group of neuropsychologists and neuroscientists in California.[2] She is known by the initials A. J. and was aged thirty-four when first tested in 2000. Her special abilities are restricted to autobiographical memory—memory for daily events that she experienced personally—but whereas most people may remember events of the last few days or those surrounding unusual happenings, A. J. can remember every day since she was nine years old. When given a random date (say, April 3, 1990), she is able to say what day of the week it was, what she was doing (e.g., going out to a named restaurant with a named friend), and give details of any significant public events that occurred on that day. Her memories are deeply personal, vivid, and full of emotion. She reported that for her, remembering is automatic and not under conscious control; it also seems that one recollection triggers another in an associative manner. For some years, A. J. had kept a diary, so the researchers were able to verify the details she provided. Like the mnemonist S, A. J.'s unusual abilities were experienced more as a handicap than as a boon or blessing; she admitted to being bothered by being automatically reminded of many past personal experiences.

Interestingly, A. J. did not score well on many clinical tests of cognitive function. Her IQ was in the normal range; she scored well on verbal memory tests, but did poorly on memory for faces and abstract drawings. She has difficulty with rote memorization, was not good at school, and has to make lists to remember things! The researchers noted that A. J. spends an abnormal amount of time thinking about her past, and a later investigator observed aspects of obsessive-compulsive behavior in her personality. She was given a brain scan, but both her hippocampus and prefrontal cortex were reportedly normal. As more of these cases of extraordinary memory come to light, it should be possible to link their unusual abilities to unusual features of brain anatomy or brain function, but for the moment we may remark that cases can be as different as S and A. J.

Expertise as a Basis for Excellent Memory

So far in this chapter, we have talked about cases of excellent memory that appear to reflect idiosyncratic differences in the brains of the people showing these strikingly unusual abilities. As far as is known at present, the differences are at least in part due to the inborn "hardware" of the brains in question and are not fully attributable to any actions taken by the people themselves. There *are* actions that can be taken by ordinary individuals to improve (or maintain) good brain functioning with positive effects on memory, and we discuss such actions in a later section. But for now we will turn to memory improvements that follow from deliberate strategies employed by individuals, or from knowledge, skills, and expertise acquired in the course of daily living, training, and learning.

In chapter 2, we described the levels-of-processing framework, and illustrated how processing words deeply and meaningfully resulted in successful later recollection. We can now make a more general point by stating that the key to much excellent memory involves relating the items to be remembered to some highly organized body of relevant knowledge that the person possesses. These bodies of knowledge provide *schematic support*, and can exist in many forms depending on the person's personal as well as professional interests and pursuits. Everyone who is not visually impaired is an "expert" in visual perception given that we have used this faculty from an early age to interpret the visual world, and this is probably the reason that memory for pictures is so outstanding (see chapter 3).

The finding from everyday observation, and some studies, is that expertise supports excellent memory for events, facts, and actions related to the expertise in question. As an example, Dutch psychologist Adriaan de Groot carried out a particularly illuminating study of memory for chess positions.[3] He placed twenty different chess pieces on a board, simulating a game in progress, and asked players ranging from amateurs to grand masters to study the board briefly and then replace the pieces in their correct locations from memory. Unsurprisingly, the grand masters easily outperformed the amateurs on this task; the masters were about 93 percent correct in their placements, whereas the amateurs placed only 50 percent correctly. A possible conclusion is that chess grand masters may possess a form of photographic memory; perhaps this ability even enabled them to become grand masters. De Groot gave the players a second test, however, this time placing the twenty pieces in random locations on the board. Now the grand masters were no better than the amateurs in remembering the placements, thereby disconfirming the "photographic theory," but demonstrating that their chess expertise enabled them to rapidly perceive and encode meaning from the genuine board settings, and so remember them later.

Constructing Schemata as Memory Aids

It may not seem too surprising that experts can remember many pieces of information from their own knowledge base, even though the information is an arbitrary selection from their extensive repertory. The same principle, however, may be used to construct a meaningful framework that can be used to remember unrelated items, provided that the items can be associated meaningfully with some aspect of the framework. A simple example that supplies a means for remembering ten items in the order they were presented is given by the children's rhyming scheme "One is a bun, two is a shoe, three is a tree, four is a door, five is a hive, six is sticks, seven is heaven, eight is a gate, nine is wine, ten is a hen." To use this memory aid (or *mnemonic device*), a person must first memorize the rhyming sequence and then associate each object in turn with the next object in the rhyme. So if the list to be remembered starts with "lettuce, potatoes, eggs, milk . . ." the person might construct images of enjoying lettuce in a bun sandwich, a potato stuck in a shoe, making walking difficult, eggs fastened to a tree, bottles of milk standing outside the person's front door, and so on. Without such an aid, it is quite difficult to remember ten unrelated objects in their correct serial order, but with the rhyming aid it is remarkably easy. Usually it is difficult to associate wanted items to the rhyming objects in a rational way (eggs don't grow on trees, for example!), but bizarre or surreal images are at least as effective; also, surprisingly perhaps, the rhyming scheme can be reused many times without much interference from previous occasions. Armed with this trick, you can amaze your friends at the next (suitably boring!) dinner party.

This general strategy of mentally associating items to be remembered with aspects of well-learned structured knowledge has actually been known and practiced for at least two thousand years. It was known to ancient Greek and Roman orators, who would associate various arguments they wished to make in a speech with successive memorized locations in their house or street. These locations ("loci" in Latin) then served as "points" to make in their speeches, which were given without notes. The strategy is now known as the *method of loci* and may take several forms, one of which is for the person to memorize an organized walk around their house or town until they can effortlessly envisage the locations and their order. Once learned, the sequence of locations can then be used as "pegs" on which to hang items to be remembered, using vivid mental associations for each item. The information can then be retrieved in order by mentally retracing the learned route.

Variants of this method are typically used by people who compete in memory championships. These "memory champions" can perform amazing feats such as memorizing lists of over three hundred random digits in five minutes and recalling the sequence of a deck of fifty-two playing cards after studying the deck for thirty seconds. They do this by first learning an elaborately structured series of mental locations—one champion constructed a "memory palace" containing three hundred sequential locations!—and then associating each of the digits, or playing cards, with each successive location. They may also recode the numbers zero through nine and each of the fifty-two playing cards into distinctive imaged objects to facilitate their association with the various loci in the memory palace or memory walk. A further practiced skill is the ability to make distinctive associations in a short time.

This general strategy of mentally associating items to be remembered with aspects of well-learned structured knowledge has actually been known and practiced for at least two thousand years.

Memory researchers have been intrigued by the abilities of memory champions and wondered whether their brains are abnormal in some way that enables them to perform their amazing feats. Yet some British neuroscientists have compared groups of champions and normal control individuals, and found that the champions' superior abilities were not driven by exceptional intellectual abilities or structural brain differences.[4] The study did find that while learning and retrieving different types of material, the champions used areas of the brain associated with spatial navigation, especially the posterior region of the right hippocampus. This finding is in line with the champions' reports that they were using the method of loci to remember the material. But unlike the London taxi drivers mentioned in chapter 7, the right hippocampus of memory champions was not enlarged, possibly because the champions simply used the hippocampus during their feats of memorization rather than modifying the structure to store vast amounts of spatial knowledge. One last point about memory champions; in line with our description of experts using acquired schemata to memorize and retrieve various types of information, the champions' abilities depend on specialized strategic skills they can use to deliberately remember long strings of numbers and the order of cards in a deck, but they are no better than anyone else at remembering where they left their glasses or what they came upstairs for! Their exceptional memories reflect a specialized skill and not a superior memory for everyday events.

Maintaining Good Memory

In the case of biological, brain-based differences between individuals, factors that may improve memory by making the brain function more effectively include diet, exercise, sleep, and certain drugs. These variables are likely to have general, nonspecific effects on both brain and memory functions, and may improve aspects of memory through increased alertness along with the enhanced effectiveness of encoding and retrieval operations. We discuss such factors in the following sections.

Diet. According to the US Department of Health and Services, more than two-thirds of adults are overweight or obese, largely due to poor diet. A culprit in this level of obesity is a reliance on fast foods, which are loaded with food additives along with unsaturated fats. Further, a common supposedly "good meal" is a steak and baked potato covered with butter and sour cream, followed by a piece of pie covered with ice cream. Excessive consumption of a diet of these sorts is associated not only with obesity but also with cognitive decline and dementia resulting from neuronal alterations in the hippocampus and prefrontal cortex brain regions essential for encoding memories and controlling behavior. This was found in a study showing that women who ate the most saturated fats from foods such as red meat and butter performed worse on tests of thinking and memory than did women who ate the lowest amounts of these fats.

The buildup of cholesterol plaques in brain blood vessels can damage brain tissue, resulting in brain cells being deprived of the oxygen-rich blood needed to function normally. Doing so compromises thinking and memory. The reason for the connection between diets high in saturated and trans fats and poorer memory may be mediated by a gene called apolipoprotein E (APOE). This gene is associated with the amount of cholesterol in your blood. People with a variation of this gene, called APOE e4, are at a greater risk for Alzheimer's disease. Those with this genetic variation have a greater number of sticky protein clumps, called beta-amyloid plaques, in the brain. These plaque deposits are associated with the destruction of brain cells and are a hallmark of Alzheimer's disease, although whether such plaques are *responsible* for the symptoms of Alzheimer's is still under debate.

Fruits, vegetables, whole grains, fish, and olive oil help improve the health of blood vessels, reducing the risk for a memory-damaging stroke. Fish are high in omega-3 fatty acids, which have been linked to lower levels of beta-amyloid proteins in the blood and better vascular health. Some will be pleased to learn that moderate alcohol

consumption raises levels of healthy, high-density lipopro-
tein cholesterol. Alcohol also lowers our cells' resistance to
insulin, allowing it to lower blood sugar more effectively.
Insulin resistance has been linked to dementia. Of course,
an excessive use of alcohol is bad for memory, potentially
leading to Korsakoff syndrome, a form of amnesia.

Sleep. The growth of obesity over the last thirty years
parallels a growth in self-reported sleep deprivation, suggesting a link between the two. Sleep deprivation interferes
with attention, leading to a decline in cognitive control. In
turn, attentional control is important for encoding and
retrieval, leaving individuals prone to errors reflecting an
overreliance on automatic influences of memory. A study
reported that getting even one hour of sleep less than the
optimal level (seven to eight hours nightly) leads to problems with concentration, attention, and memory. Sleep
deprivation increases the risk of Alzheimer's disease too.
Sleep and the internal circadian clock influence a host of
endocrine parameters leading to insulin resistance. Additionally, low levels of beta-amyloid 42 in the cerebrospinal
fluid signify the presence of amyloid plaques, related to
Alzheimer's disease.

Older research focused on the role of rapid eye movement (REM) sleep; recent work has shown the importance
of slow-wave sleep (SWS) for memory consolidation. Newer
findings characterize sleep as a brain state that is optimal
for memory consolidation, in contrast to the waking brain

being optimized for the encoding and retrieval of memories. As described in chapter 7, consolidation reflects the reactivation of recently encoded memories and integrates representations into long-term memory. Reconsolidation is said to occur during SWS, whereas REM sleep may stabilize transformed memories. An excellent account of the effects of sleep on memory and the brain is given by Matthew Walker and Robert Stickgold.[5]

Exercise. Memory and thinking are indirectly helped through exercise by reducing obesity and increasing sleep as well as reducing the negative consequences for memory produced by stress and anxiety. More directly, exercise enhances memory by its ability to reduce insulin resistance, reduce inflammation, and stimulate the release of chemicals in the brain that affect the health of brain cells, growth of new blood vessels in the brain, and even abundance and survival of new brain cells as well as cardiovascular function. The prefrontal and medial temporal cortices are brain areas that are crucial for attention and memory; they have greater volume in people who exercise versus people who do not.

Aerobic exercises are most beneficial for enhancing memory and cognition. Examples of aerobic exercises include jogging and walking as well as other strenuous activities. The best measure of cardiovascular function is maximal oxygen consumption during strenuous exercise. Older adults, just like young ones, show significant benefits in maximal oxygen consumption following prolonged endurance training. Cardiovascular exercises increase the stroke volume of the heart and oxygen transportation to the brain. Research shows that cardiovascular exercise significantly improves oxygen flow to the brain. Correlational studies compare the cognitive performance of people who exercise regularly to those who do not. The results from such studies show that exercisers perform better on a variety of tasks, including the ability to learn and recall a list of unrelated words as well as working memory tasks such as the ability to hold items in mind over brief periods of time when distracted. One caveat is that the results demonstrating a correlation between exercise and memory are open to the possibility that the finding was produced by a factor other than exercise. For example, those who exercise may have better overall health than those who do not, with the correlation reflecting general levels of health rather than amounts of exercise.

More direct evidence of the benefits of exercise comes from studies in which engagement in exercise is directly manipulated.[6] In such experiments, the participants in one condition are assigned to a group that engages in exercise such as brisk walking for a fixed amount of time on a schedule that extends over a few months. The results from memory tasks performed by the participants in the exercise condition are compared to the results from the participants in control conditions in which vigorous exercise is not involved. Such experiments have revealed that the participants in the exercise condition produce higher performance on memory tasks as well as other tests of cognition than do those in a control, nonexercise condition. The form of exercise is important, however, with the advantages in memory associated only with aerobic exercise and not with nonstrenuous forms of exercise. Older adults benefit from exercise just as do young adults. Yet the benefits of exercise differ across memory tasks used to measure its effects; for example, some studies show that the benefits of exercise are greater for memory tasks that involve the frontal lobes. This is significant in that as described in prior chapters, the frontal lobes are the brain regions that show the greatest deterioration as a function of age.

In sum, the results offered in this section strongly suggest that to maintain or improve memory and other cognitive functions, one should adopt a healthy diet, get an adequate amount of sleep, and engage in strenuous exercise. Doing so is not only beneficial for memory and cognitive functions but also increases longevity. Having followed these recommendations, one can reward oneself by having an occasional alcoholic beverage, which is both beneficial and pleasurable!

Cognitive enhancers. Advertisements shown on television promote the use of special compounds that are said to greatly improve memory performance. Whereas the focus of many research studies has been on improving cognitive performance in patients with brain disorders, these advertisements feature compounds that are claimed to enhance memory in healthy individuals. The ads often include endorsements from people who have benefited from taking the compounds. There are reasons to be skeptical about such testimonials, though. More objective means of evaluating the effectiveness of dietary supplements for improving memory lead to a less optimistic conclusion. One review of such compounds concludes that the drugs have relatively modest effects that vary substantially between and even within individuals.[7] The review also looks at undesirable side effects associated with the compounds. Another review ends with the author naming a favorite drug—1,3,7-trimethylxanthine—that enhances many aspects of cognition by improving alertness. The drug also goes by another name: caffeine!

Memory Rehabilitation

In the initial sections of this chapter portraying cases of amazing memory, we mentioned that some instances reflect expertise in a domain, as in that of London taxi drivers. The effects of preserved expertise are also shown in high memory performance in older adults. For example, expert players in the card game Bridge can better remember hands of cards played earlier in a game than can nonexperts. The effects of expertise are shown in more everyday memory activities too. In this vein, one of the authors (Jacoby) talked with a neighboring older adult about the difficulty of remembering names. The neighbor responded that she never experienced such difficulty, saying that she had a career in sales for which memory for names was important. Such observations suggest that it is possible to improve memory functioning by training specific cognitive operations like spaced retrieval, face-name associations, elaborative encoding, and organization along with the use of a person's expert knowledge to support more efficient encoding and retrieval operations. These cognitive strategies work for rehabilitating memory as well, as we will illustrate, but the effects tend to be restricted to specific materials and situations just as is found in memory experts. Typically they do not provide a general boost to all memory abilities.

A successful memory rehabilitation procedure aimed at improving the memory of healthy older adults should fulfill three criteria. First, it should substantially increase performance in the targeted area. For instance, if the program is aimed at using the method of loci, it should result in the participants learning to successfully use that procedure. Second, a successful procedure should transfer beyond the specific materials and tasks trained. In the context of memory rehabilitation, "near transfer" indicates some transfer to materials and tasks that are similar to those used for training. "Intermediate transfer" refers to gains in other forms of memory following training, and "far transfer" refers to gains in other cognitive domains, particularly those in everyday life. Third, a successful intervention should result in transfer that persists over a substantial period—such as months rather than minutes. Many training procedures have shown improvements on the tasks used for training, and that the improvements persist over long periods. Fewer procedures have demonstrated appreciable amounts of intermediate transfer, and findings of far transfer have been rare.

A hallmark feature of cognitive aging is a failure to *self-initiate* the use of memory strategies. One approach to memory rehabilitation has been to train self-initiated strategies such as semantic elaboration. Importantly, older adults can be trained to use such strategies as shown by results in word list learning, face-name learning, and working memory tasks. Also, older adults can learn to use the method of loci, a mnemonic strategy used by some memory experts, as discussed earlier in this chapter. Older adults trained to use that strategy show impressive memory performance. Still, as noted for the mnemonists, the transfer to other tasks is extremely limited. Most older adults do not continue to use the strategy after training, perhaps because it is impractical for everyday activities. Most training studies have involved only a single strategy. In contrast, one large study trained older adults to use a variety of strategies, not only in laboratory tasks, but also in everyday-like ones.[8] The results revealed impressively large effects that lasted as long as five years. The primary finding, however, was that the effects were specific to the domain of training. Not only are older adults less likely to spontaneously use mnemonic strategies until trained to do so, the training of mnemonic strategies is often found to magnify age differences in memory performance by showing that young participants gain more from such training than do older adults. That is, while such training can certainly be beneficial for older adults, it does not necessarily result in the elimination of age differences.

Training Recollection

A promising alternative to training memory strategies is to train underlying memory processes, such as training for a reliance on recollection, thereby reducing errors resulting from a more automatic use of memory. Janine Jennings and Larry Jacoby developed a procedure for doing so, and found encouraging results.[9] The training procedure was patterned after discouraging errors of the sort made by the confused professor who flew home from a conference to which he had driven, described in chapter 4. The procedure involves a study phase followed by a test phase during which studied words are intermixed with new words with

A hallmark feature of cognitive aging is a failure to *self-initiate* the use of memory strategies.

some new words being repeated after varying delays. The participants were warned that some new words were repeated and they should respond "new" to such words. This is an opposition paradigm in that mistakenly responding "old" to repeated new words would result from responding on the basis of familiarity, akin to flying home from a conference to which one had driven. In contrast, responding "new" to repeated new words would result from a reliance on recollection. Gradually increasing the delay between repetitions of new words was meant to train recollection under increasingly difficult conditions. The results from such experiments have been impressive, showing that healthy older adults and even participants in the early stages of Alzheimer's disease improved their ability to reject repeated new words after long delays. Intermediate transfer, including transfer to a working memory task, has also been found. Nicole Anderson has used a variant of the technique and demonstrated that improvements were maintained for three months.[10] Further, estimates of the separate contributions to the performance of recollection and familiarity were computed using methods discussed in chapter 4. The effects of training were shown to result from an increase in the contribution of recollection, with that of familiarity unchanged. This result is the same as those reported in chapter 4 comparing the performance of younger to older adults and full attention to divided attention.

It is generally accepted that maintaining an active lifestyle in terms of social interactions and challenging cognitive activities (e.g., doing crosswords and sudokus) is the way to maintain a good level of cognition at age sixty and beyond. Various studies have explored this assumption experimentally—one by contrasting the effects of various activities on the cognitive performance of older adults.[11] Some activities were classified as involving "productive engagement." For example, one group was given instruction and practice in digital photography including photo editing, a second group had instruction and practice in quilt making, and a third group spent half of its time learning photography and half making quilts. Each group worked for an average of fifteen hours per week over fourteen weeks. These groups were contrasted with two further groups whose members performed activities classified as "receptive engagement." One group met to perform social activities such as watching movies and going on trips; a second group read magazines and listened to classical CDs; and a final group was given no treatment. Before and after performing these various activities, all the groups were given a battery of cognitive tests. The main result was that the productive engagement groups showed substantial improvement on a test of episodic memory; this was largely attributable to the photo group, although the quilting group also improved relative to the no-treatment group. The receptive engagement groups showed no improvement. A final result of the study was that the productive engagement groups also showed evidence of increased neural efficiency in certain brain regions (using fMRI), and some of the increased efficiency was maintained a year later. The authors commented that the results illustrate the benefits of keeping the mind engaged and learning new skills, and provide support for the "use it or lose it" hypothesis of cognitive aging.

Training Working Memory

Recently, computer games have been devised as a means of maintaining and rehabilitating memory. Such games are attractive in that they are readily available and free the participants from the constraints imposed by laboratory training procedures. Also, they are fun to play. Studies examining the effectiveness of computer training have been encouraging, but have produced mixed results, with some showing limited gains and little transfer to other situations. One apparent exception is a study using *NeuroRacer*, a computer game in which players attempt to steer a car along a winding road while also responding rapidly to occasional road signs, but only when a green circle is present.[12] Older adults were trained on increasingly difficult versions of the game for twelve one-hour sessions over the course of a month. The training greatly improved their game performance as well as enhanced their performance on other laboratory tasks involving working memory and sustained attention—a benefit that lasted at least six months. A further important finding using an EEG recording was that the neural connectivity between the frontal lobes and posterior brain regions was enhanced by the training. The authors stress the point that the game involves multitasking—simultaneously performing the sign task while maintaining the car in the center of the winding road. The study thus demonstrates intermediate transfer (i.e., to other laboratory tasks), but it is still unclear whether the benefits would be shown in real-life driving situations too.

A pervasive problem for attempts to rehabilitate memory has been the finding of limited transfer across tasks. The above study provides an illustration of a promising means of dealing with that problem by jointly investigating both the behavioral and neural consequences of training. Finding a lack of transfer between tasks is hardly surprising if the tasks involve different brain areas. In this regard, it has been found that brain areas involved in a task differ between young and older adults, sometimes being restricted to a single frontal lobe for young adults while involving bilateral frontal activation for older adults. Such differences suggest the use of compensatory strategies by older adults. Examining behavioral differences in combination with brain differences potentially provides a means of understanding the bases for transfer across tasks. Cindy Lustig and colleagues have pulled together an excellent review of results from studies including both the behavioral and neural consequences of memory training.[13]

Failing Memory? There's an App for That!

In our view, the memory deficits associated with normal healthy aging are irritating but not critical. Failures to remember words and names are annoying, and sometimes embarrassing, but they do not really interfere much with life. Further, memory problems are lessened by changes in the demands made of memory. The invention of the printing press in the fifteenth century along with increases in general literacy greatly changed the role that memory played in life. Contracts could now be produced in a printed form, for instance, rather than being left to the memory of those involved. Currently, access to computers along with increasing computer literacy is having a corresponding effect on memory for facts, as in simply Google it! Reading a book with interleaving characters and plots can be difficult, but the problem can be solved with a Kindle book that provides search programs. Cell phones provide a convenient and effective way to remember appointments and other things to be done in the future. Has this increasing reliance on technology affected our ability to remember?

This is an interesting question that has not yet been answered definitively by research studies. It seems clear that the *use* of memory is less crucial in modern life; politicians use the teleprompter instead of the method of loci when making speeches! But there is little reason to believe that our *capacity* to learn and remember names, dates, facts, and events has declined. The underlying biology is still intact; we just have to use it more effectively.

We hope that this brief survey of human memory from the point of view of cognitive psychology has answered at least some of the questions you had about your own memory experiences. We have emphasized the point that good memory is often tied to meaning and expertise. Knowledge in a certain area provides a framework for facts and events in that area, but does not improve memory and learning generally. Mnemonic techniques, such as the one-is-a-bun rhyme and method of loci, involve creating such frameworks at a personal level. The framework can then be used as an organized series of pegs on which to hang items to be remembered. This is the technique frequently used by memory champions to perform astonishing feats of remembering—but as we mentioned, they still forget where they left their car keys!

Ways to improve memory more generally include attention to diet, sleep, and especially exercise; there is now good evidence that exercise is associated with improved cognition, possibly attributable to better blood supply to the brain and the generation of new brain cells. We also described some promising recent research directed at improving working memory (the *NeuroRacer* computer game), and improving recollection by sharpening the distinction

Ways to improve memory more generally include attention to diet, sleep, and especially exercise.

between old words presented in an initial list and repeated new words. Other effective methods include retrieving names and facts at progressively longer-spaced intervals, constructing meaningful organizations for items to be remembered, and meaningful associations between faces and names. We are somewhat skeptical about the value of smart drugs and cognitive brain games at the present time, although both methods attract a great deal of interest. Future research efforts may yet lead to products with less equivocal results, but for the moment there is no magic pill.

In terms of theoretical approaches to understanding memory more fully, we stressed the importance of attentional control, meaningfulness, and organization in the creation of good episodic memories. We also mentioned the crucial notion of consolidation in chapter 7—the idea that a cascade of neural processes links the hippocampus to the cerebral cortex as a means of fixing perceived events as semipermanent memories. When it comes to retrieving memories, we underscored the central importance of reestablishing the context in which an event originally occurred. There is now good evidence from studies using fMRI that the pattern of brain activation seen during the encoding of an event is largely seen again in the course of successful retrieval. The outside environment can often help to reinstate the relevant context, or else cognitive control processes originating in the frontal lobes must be recruited to drive the brain back into the appropriate configuration.

We also described how cases of impaired memory (notably that of patient H. M.) can lead to critical insights into the factors associated with the strengths and weaknesses of memory. In H. M.'s case, this led to the distinction between the ability to learn habitual procedures (which H. M. possessed) and the ability to recollect the occasion of learning (which he lacked). In addition, investigations of healthy aging have yielded useful information about what holds up well (e.g., holding a telephone number in mind, and the processes associated with feelings of familiarity) and what does not (e.g., the ability to consciously recollect details of experienced events).

In summary, the combined efforts of anatomists, biochemists, physiologists, psychologists, and neurologists have led to many important findings and theories relating to the structures as well as processes involved in human memory experience and performance. Much more remains to be discovered, however. We hope that at least we have interested you in these many remaining puzzles and loose ends. If so, in the further reading chapter we recommend a few accessible sources that should provide satisfying answers and (more important) suggest some further fascinating questions.

GLOSSARY

Amygdala
The amygdalas are a pair of almond-shaped structures located deep in the medial-temporal lobes of the brain (figure 6); they play a primary role in the processing of emotional information.

Anagram
Anagrams are two words using the same rearranged letters. Examples include "secure"/"rescue" and "carthorse"/"orchestra." Solving an anagram (finding the alternative word) requires effortful and often semantic processing, with positive effects on later memory (chapter 4).

Attention
The essential characteristics of attention in the context of this book are its selectivity, limited capacity, and role in working memory and cognitive control.

Attributions
The interpretation of a pattern of neural activation is sometimes attributed to the wrong source. For instance, ease of reading may be attributed to previous knowledge (chapter 4).

Cerebral cortex
The cerebral cortex is the outer layer of the mammalian brain (figure 6); it is associated with higher mental functions.

Conscious recollection
This term refers to the conscious remembering of a previously experienced event, including its time and place of occurrence.

Consolidation
The neural processes by which perceived experiences are stabilized in the brain to form long-term memories.

Context reinstatement
Remembering an experienced event fully involves a recollection of the context in which it occurred. A re-presentation of the original context helps to reconstruct the memory.

Control processes
The brain processes associated with the frontal lobes that manage and support cognitive processing to overcome habitual responding, and fulfill some current plan or goal.

Dissociation
As used in this book, dissociation refers to the situation in which two aspects of an ability are affected differently by some condition or manipulation. For example, the processes underlying habit and recollection, two aspects of memory, are affected differently by aging. Older age is associated with a decrease in recollection, but habit is unchanged by aging (chapter 4).

Dual process account
This phrase refers to the idea that recognition memory depends on both general familiarity and conscious recollection of some specific earlier experience.

EEG
EEG stands for electroencephalography, a technique for measuring and recording the electric activity of brain processes as they occur.

Encoding
The processes involved in the perception and consolidation of memories in the brain.

Environmental support
This is the idea that remembering is driven partly by the self-initiated reactivation of stored memory traces, but also by information from the external environment. The two sources provide complementary information (chapter 6).

Episodic memory
Episodic memory is the conscious recollection of a specific previously experienced personal event; the concept is also referred to as autobiographical memory.

Executive functions
A set of brain processes (also known as control processes) that serve to regulate and manage memory, cognition, and action. They are mediated by the frontal lobes, and include attention, inhibition, and working memory.

Flashbulb memories
These are the long-lasting vivid memories of some emotionally charged event (chapter 5).

fMRI
This acronym stands for functional magnetic resonance imaging (see also MRI). It is a brain-scanning technique that measures the small changes in blood supply that accompany brain processes. The measures are typically recorded while the participant is engaged in some form of cognitive activity.

Forgetting curve
In 1885, Hermann Ebbinghaus published one of the first books reporting experimental work on human memory. He discovered that forgetting occurred rapidly at first after learning ceased, but then progressed more slowly as further time elapsed.

Free recall
Free recall is an experimental technique in which the participant first studies a set of words, pictures, or objects, and later attempts to recall them in any order without aids or prompts.

Gestalt psychology
The school of gestalt psychology flourished in the first half of the twentieth century. The movement stressed that perception and other mental processes must be understood in terms of whole patterns rather than as built up from fundamental elements.

Habitual influences
Well-learned aspects of knowledge and behavior can affect our current thinking and actions in the absence of conscious intention or awareness. In chapter 4, we illustrate how habitual influences and conscious recollections are sometimes opposed, and sometimes act in concert.

Hippocampus

The hippocampus is a complex structure in the medial temporal lobe of the brain (figure 6). It plays a major role in navigation and the consolidation of memory in humans and other mammals.

Implicit memory

This term refers to stored information that is not available to consciousness but does affect thought and action. Other terms for the same concept are "unconscious influences" and "memory without awareness."

Involuntary memories

Involuntary memories are those that occur without any conscious intention to remember. They are typically invoked by an environmental pattern that formed a salient part of some earlier event (chapter 5).

Levels of processing

The levels-of-processing framework for memory research stresses that remembering is an active process of mind. The initial experiments focused on the different consequences for later memory when encoding processes dealt with surface details (shallow processing) as opposed to meaningful aspects (deep processing) of perceived events.

Long-term memory

Long-term memory refers to the stable, semipermanent storage of information no longer in conscious awareness. Some of the information can be recalled consciously (explicit memory), whereas other information can affect behavior unconsciously (implicit memory).

Memory systems

The term refers to the idea that memory exists in different forms that have different characteristics and obey different rules. Explicit memory includes working, episodic, and semantic memory; implicit memory includes learned skills and procedures.

Mental time travel

This is a term emerging from Endel Tulving's concept of episodic memory. It refers to the ability to travel back in time mentally to envisage a previously experienced personal event.

Metacognition
The awareness and understanding of one's own cognitive processes, including thinking, learning, and memory.

Method of loci
A mnemonic device in which a person uses the mental images of a prelearned series of places (loci in Latin), such as rooms in a house, as mental pegs on which to hang the images of the items to be remembered.

MRI
Magnetic resonance imaging is used in neuroscience to obtain images of brain structures and their connections (see also fMRI).

Neuroimaging techniques
These techniques are used to obtain information about the brain's structures and functions. The methods commonly used in studies of memory include EEG, fMRI, and PET (positron emission tomography).

Organization
In this book, organization refers to the mental grouping of words or other items into meaningful categories or structures. In turn, such organization benefits later recollection.

Proactive interference
This refers to interference with current learning arising from previously learned material of a similar type.

Procedural memory
This term refers to the (usually unconscious) retention of learned mental skills (e.g., reading) and physical skills (e.g., skating).

Processing resources
This is the idea that cognitive processes such as thinking and remembering require some form of mental energy to function effectively. Such resources are often attributed to our limited attentional capacity.

Prospective memory
Prospective memory is about remembering to do things in the future after planning to do so (chapter 6).

Repression
Repression is a Freudian term referring to the mind's ability to block painful and traumatic memories from conscious awareness.

Retroactive interference
This term refers to the interference of current learning by subsequent information of a similar type.

Semantic memory
Semantic memory is one of Endel Tulving's proposed memory systems. It refers to general knowledge that we possess in the absence of knowing where or when we learned it.

Short-term memory
This term is used popularly to refer to memory for recent events. This type of memory is vulnerable to a variety of conditions (e.g., normal aging or mild cognitive impairment) that leave memory for remote events intact. In Richard Atkinson and Richard Shiffrin's model, it has a more restricted meaning: the small amount of information held in conscious awareness.

Spacing effect
This is the finding that learning shows a benefit when repeated lessons or exposures to learned material are spaced widely apart as opposed to being massed together.

State-dependent memory
If information is learned when a person is in a given mental state or physical context, remembering is more effective when the original state or context is reinstated.

Testing effect
This is the finding that the retrieval of recently learned material is more effective for subsequent remembering than is a second learning opportunity.

Tip-of-the-tongue effect (TOT)
This is the experience of being unable to recollect some name or other information, accompanied by the strong feeling that the information is known and *almost* retrievable (chapter 3).

Unconscious influences

This term refers to the finding that previously experienced events or learned material can affect present mental or physical behavior without the person being aware of that influence (chapter 4).

Working memory

Working memory refers to the small amount of information that can be held temporarily in conscious awareness. The information is worked on by cognitive processes to participate in learning, reasoning, and decision-making.

NOTES

Chapter 1
1. W. B. Scoville and B. Milner, "Loss of Recent Memory after Bilateral Hippocampal Lesions," *Journal of Neurological and Neurosurgical Psychiatry* 20 (1957): 11–12.
2. F. C. Bartlett, *Remembering: A Study in Experimental and Social Psychology* (Cambridge: Cambridge University Press, 1932).
3. A. D. Baddeley, "Short-Term Memory for Word Sequences as a Function of Acoustic, Semantic, and Formal Similarity," *Quarterly Journal of Experimental Psychology* 18 (1966): 302–309; A. D. Baddeley, "The Influence of Acoustic and Semantic Similarity on Long-Term Memory for Word Sequences," *Quarterly Journal of Experimental Psychology* 18 (1966): 362–365.
4. E. Tulving and D. L. Schacter, "Priming and Human Memory Systems," *Science* 247 (1990): 301–306.
5. R. S. Rosenbaum, S. Köhler, D. L. Schacter, M. Moscovitch, R. Westmacott, S. E. Black, F. Gao, and E. Tulving, "The Case of K.C.: Contributions of a Memory-Impaired Person to Memory Theory," *Neuropsychologia* 43 (2005): 989–1021.

Chapter 2
1. R. C. Atkinson and R. M. Shiffrin, "The Control of Short-Term Memory," *Scientific American* 225 (1971): 82–90.
2. G. A. Miller, "The Magical Number Seven, Plus or Minus Two: Some Limits on Our Capacity for Processing Information," *Psychological Review* 63 (1956): 81–97.
3. A. D. Baddeley and G. J. Hitch, "Working Memory," in *The Psychology of Learning and Motivation*, ed. G. H. Bower (New York: Academic Press, 1974), 47–89.
4. A. D. Baddeley, "The Episodic Buffer: A New Component of Working Memory?," *Trends in Cognitive Sciences* 4 (2000): 417–423.
5. N. Cowan, "An Embedded-Processes Model of Working Memory," in *Models of Working Memory*, ed. A. Miyake and P. Shah (New York: Cambridge University Press, 1999), 62–101.

6. F. I. M. Craik and R. S. Lockhart, "Levels of Processing: A Framework for Memory Research," *Journal of Verbal Learning and Verbal Behavior* 11 (1972): 671–684.

7. A. M. Treisman, "Strategies and Models of Selective Attention," *Psychological Review* 76 (1969): 282–299.

8. R. R. Hunt and J. B. Worthen, eds., *Distinctiveness and Memory* (New York: Oxford University Press, 2006).

9. F. I. M. Craik and E. Tulving, "Depth of Processing and the Retention of Words in Episodic Memory," *Journal of Experimental Psychology: General* 104 (1975): 268–294.

10. E. Tulving, "Cue-Dependent Forgetting," *American Scientist* 62 (1974): 74–82.

Chapter 3

1. J. S. Bruner, "On Perceptual Readiness," *Psychological Review* 64 (1957): 123–152.

2. R. S. Nickerson and M. J. Adams, "Long-Term Memory for a Common Object," *Cognitive Psychology* 11 (1979): 287–307.

3. D. J. Simons and D. T. Levin, "Failure to Detect Changes to People during a Real-World Interaction," *Psychological Bulletin & Review* 5 (1998): 644–649.

4. L. Standing, "Learning 10,000 Pictures," *Quarterly Journal of Experimental Psychology* 25 (1973): 207–222.

5. H. P. Bahrick, P. O. Bahrick, and R. P. Wittlinger, "Fifty Years of Memory for Names and Faces: A Cross-Sectional Approach," *Journal of Experimental Psychology: General* 104 (1975): 54–75.

6. J. W. Pichert and R. C. Anderson, "Taking Different Perspectives on a Story," *Journal of Educational Psychology* 69 (1977): 309–315.

7. R. Brown and D. McNeill, "The 'Tip of the Tongue' Phenomenon," *Journal of Verbal Learning and Verbal Behavior* 5 (1966): 325–337.

8. D. R. Godden and A. D. Baddeley, "Context-Dependent Memory in Two Natural Environments: On Land and Underwater," *British Journal of Psychology* 66 (1975): 325–331.

9. D. W. Goodwin, B. Powell, D. Bremer, H. Hoine, and J. Stern, "Alcohol and Recall: State Dependent Effects in Man," *Science* 163 (1969): 1358–1360.

10. E. Eich, "Theoretical Issues in State Dependent Memory," in *Varieties of Memory and Consciousness: Essays in Honour of Endel Tulving*, ed. H. L. Roediger III and F. I. M. Craik (Hillsdale, NJ: Lawrence Erlbaum Associates, 1989), 331–354.

11. S. M. Smith, "Remembering in and out of Context," *Journal of Experimental Psychology: Human Learning and Memory* 5 (1979): 460–471.

12. H. P. Bahrick, L. E. Bahrick, A. S. Bahrick, and P. E. Bahrick, "Maintenance of Foreign Language Vocabulary and the Spacing Effect," *Psychological Science* 4 (1993): 316–321.

13. D. L. Hintzman, "Memory Strength and Recency Judgments," *Psychonomic Bulletin and Review* 12 (2005): 858–864.

14. See, for example, L. L. Jacoby and C. N. Wahlheim, "On the Importance of Looking Back: The Role of Recursive Remindings in Recency Judgments and Cued Recall," *Memory & Cognition* 41 (2013): 625–637.

15. H. L. Roediger III and J. D. Karpicke, "Reflections on the Resurgence of Interest in the Testing Effect," *Perspectives on Psychological Science* 13 (2018): 236–241.

16. T. K. Landauer and R. A. Bjork, "Optimal Rehearsal Patterns and Name Learning," in *Practical Aspects of Memory*, ed. M. M. Gruneberg, P. E. Morris, and R. N. Sykes (London: Academic Press, 1978), 625–632.

Chapter 4

1. E. K. Warrington and L. Weiskrantz, "A New Method of Testing Long-Term Retention with Special Reference to Amnesic Patients," *Nature* 217 (1968): 972–974.

2. L. L. Jacoby and M. Dallas, "On the Relationship between Autobiographical Memory and Perceptual Learning," *Journal of Experimental Psychology: General* 3 (1981): 306–340.

3. P. A. Kolers, "Memorial Consequences of Automatized Encoding," *Journal of Experimental Psychology: Human Learning and Memory* 1 (1975): 689–701.

4. D. L. Schacter, "Implicit Memory: History and Current Status," *Journal of Experimental Psychology: Learning, Memory and Cognition* 13 (1987): 501–518.

5. G. Mandler, "Recognizing: The Judgment of Previous Occurrence," *Psychological Review* 87 (1980): 252–271.

6. E. B. Titchener, *A Textbook of Psychology* (New York: Macmillan, 1928).

7. L. L. Jacoby and K. Whitehouse, "An Illusion of Memory: False Recognition Influenced by Unconscious Perception," *Journal of Experimental Psychology: General* 118 (1989): 126–135.

8. B. W. A. Whittlesea, L. L. Jacoby, and K. A. Girard, "Illusions of Immediate Memory: Evidence of an Attributional Basis for Feelings of Familiarity and Perceptual Quality," *Journal of Memory and Language* 29 (1990): 716–732.

9. D. Witherspoon and L. G. Allan, "The Effect of Prior Presentation on Temporal Judgments in a Perceptual Identification Task," *Memory & Cognition* 13 (1985): 103–111.

10. C. M. Kelley and L. L. Jacoby, "Adult Egocentrism: Subjective Experience versus Analytic Bases for Judgment," *Journal of Memory and Language* 35 (1996): 157–175.

11. C. M. Kelley and M. G. Rhodes, "Making Sense and Nonsense of Experience: Attributions in Memory and Judgment," *Psychology of Learning and Motivation* 41 (2002): 293–320.

12. L. L. Jacoby, C. M. Kelley, J. Brown, and J. Jasechko, "Becoming Famous Overnight: Limits on the Ability to Avoid Unconscious Influences of the Past," *Journal of Personality and Social Psychology* 56 (1989): 326–338.

13. J. M. Jennings and L. L. Jacoby, "An Opposition Procedure for Detecting Age-Related Deficits in Recollection: Telling Effects of Repetition," *Psychology and Aging* 12 (1997): 352–361.

14. J. F. Hay and L. L. Jacoby, "Separating Habit and Recollection in Young and Elderly Adults: Effects of Elaborative Processing and Distinctiveness," *Psychology and Aging* 14 (1999): 122–134.

15. J. I. Caldwell and M. E. J. Masson, "Conscious and Unconscious Influences of Memory for Object Locations," *Memory & Cognition* 29 (2001): 285–295.

16. P. R. Millar, D. A. Balota, G. B. Maddox, J. M. Duchek, A. J. Aschenbrenner, A. M. Fagan, T. L. S. Benzinger, and J. C. Morris, "Process Dissociation Analyses of Memory Changes in Healthy Aging, Preclinical, and Very Mild Alzheimer Disease: Evidence for Isolated Recollection Deficits," *Neuropsychology* 31 (2017): 708–723.

17. For a description and discussion see E. Tulving, "Memory and Consciousness," *Canadian Psychology* 26 (1985): 1–12; J. M. Gardiner and A. Richardson-Klavehn, "Remembering and Knowing," in *The Oxford Handbook of Memory*, ed. E. Tulving and F. I. M. Craik (New York: Oxford University Press, 2000): 229–244.

18. A. P. Yonelinas and L. L. Jacoby, "The Relation between Remembering and Knowing as Bases for Recognition: Effects of Size Congruency," *Journal of Memory and Language* 34 (1995): 622–643.

19. C. N. Wahlheim and L. L. Jacoby, "Remembering Change: The Critical Role of Recursive Remindings in Proactive Effects of Memory," *Memory & Cognition* 41 (2013): 1–15.

20. A. Friedman, "Framing Pictures: The Role of Knowledge in Automatized Encoding and Memory for Gist," *Journal of Experimental Psychology: General* 108 (1979): 316–355.

21. W. F. Brewer and J. C. Treyens, "Role of Schemata in Memory for Places," *Cognitive Psychology* 13 (1981): 207–230.

Chapter 5
1. D. L. Schacter, *The Seven Sins of Memory* (Boston: Houghton Mifflin, 2001).
2. E. F. Loftus, G. R. Loftus, and J. Messo, "Some Facts about 'Weapon Focus,'" *Law and Human Behavior* 11 (1987): 55–62.
3. C. Laney, H. V. Campbell, F. Heuer, and D. Reisberg, "Memory for Thematically Arousing Events," *Memory & Cognition* 32 (2004): 1149–1159.
4. T. Sharot and E. A. Phelps, "How Arousal Modulates Memory: Disentangling the Effects of Attention and Retention," *Cognitive, Affective, and Behavioral Neuroscience* 4 (2004): 294–306.
5. E. A. Kensinger and D. L. Schacter, "Memory and Emotion," in *Handbook of Emotions*, ed. L. Feldman Barrett, M. Lewis, and J. M. Haviland-Jones, 4th ed. (New York: Guilford Press, 2016), 564–578.
6. L. J. Levine and M. A. Safer, "Sources of Bias in Memory for Emotions," *Current Directions in Psychological Science* 11 (2002): 169–173.
7. F. W. Colegrove, "Individual Memories," *American Journal of Psychology* 10 (1989): 228–255.
8. R. Brown and J. Kulik, "Flashbulb Memories," *Cognition* 5 (1977): 73–99.
9. U. Neisser, "Snapshots or Benchmarks?," in *Memory Observed: Remembering in Natural Contexts*, ed. U. Neisser and I. E. Hyman Jr. (San Francisco: W. H. Freeman and Company, 1982), 43–48.
10. U. Neisser and N. Harsch, "Phantom Flashbulbs: False Recollections of Hearing the News about *Challenger*," in *Affect and Accuracy in Recall*, ed. E. Winograd and U. Neisser (New York: Cambridge University Press, 1992), 9–31.
11. W. Hirst, E. A. Phelps, R. Meksin, C. J. Vaidya, M. K. Johnson, K. J. Mitchell, R. L. Buckner, et al., "A Ten-Year Follow-up of a Study of Memory for the Attack of September 11, 2001: Flashbulb Memories and Memories for Flashbulb Events," *Journal of Experimental Psychology: General* 144 (2015): 604–623.
12. D. Berntsen, "The Unbidden Past: Involuntary Autobiographical Memories as a Basic Mode of Remembering," *Current Directions in Psychological Science* 19 (2010): 138–142.
13. N. M. Hall, A. Gjedde, and R. Kupers, "Neural Mechanisms of Voluntary and Involuntary Recall: A PET Study," *Behavioural Brain Research* 186 (2008): 261–272.
14. E. F. Loftus and J. E. Pickrell, "The Formation of False Memories," *Psychiatric Annals* 25 (1995): 720–725.

15. E. F. Loftus, "Eavesdropping on Memory," *Annual Review of Psychology* 68 (2017): 1–18.

16. J. W. Schooler, M. Bendiksen, and Z. Ambadar, "Taking the Middle Line: Can We Accommodate Both Fabricated and Recovered Memories of Sexual Abuse?," in *Recovered Memories and False Memories: Debates in Psychology*, ed. M. A. Conway (Oxford: Oxford University Press, 1997), 251–292.

17. E. F. Loftus, D. G. Miller, and H. J. Burns, "Semantic Integration of Verbal Information into a Visual Memory," *Journal of Experimental Psychology: Human Learning and Memory* 4 (1978): 19–31.

18. E. F. Loftus and J. C. Palmer, "Reconstruction of Automobile Destruction: An Example of the Interaction between Language and Memory," *Journal of Verbal Learning and Verbal Behavior* 13 (1974): 585–589.

19. D. S. Lindsay, "Misleading Suggestions Can Impair Eyewitnesses' Ability to Remember Event Details," *Journal of Experimental Psychology: Learning, Memory, and Cognition* 16 (1990): 1077–1083.

Chapter 6

1. T. A. Salthouse, "The Processing-Speed Theory of Adult Age Differences in Cognition," *Psychological Review* 103 (1996): 403–428.

2. L. Hasher and R. T. Zacks, "Automatic and Effortful Processes in Memory," *Journal of Experimental Psychology: General* 108 (1979): 356–388.

3. D. C. Park, G. Lautenschlager, T. Hedden, N. S. Davidson, and A. D. Smith, "Models of Visuospatial and Verbal Memory across the Adult Life Span," *Psychology and Aging* 17 (2002): 299–320.

4. F. I. M. Craik, E. Bialystok, S. Gillingham, and D. T. Stuss, "Alpha Span: A Measure of Working Memory," *Canadian Journal of Experimental Psychology* 72 (2018): 141–152.

5. F. I. M. Craik, "On the Transfer of Information from Temporary to Permanent Memory," *Philosophical Transactions of the Royal Society, B: Biological Sciences* 302 (1983): 341–359.

6. F. I. M. Craik and M. Byrd, "Aging and Cognitive Deficits: The Role of Attentional Resources," in *Aging and Cognitive Processes*, ed. F. I. M. Craik and S. E. Trehub (New York: Plenum Press, 1982), 191–211.

7. L. Bäckman, B. J. Small, Å Wahlin, and M. Larsson, "Cognitive Functioning in Very Old Age," in *The Handbook of Aging and Cognition*, ed. F. I. M. Craik and T. A. Salthouse, 2nd ed. (Mahwah, NJ: Lawrence Erlbaum Associates, 2000), 499–588.

8. E. Failes, M. S. Sommers, and L. L. Jacoby, "Blurring Past and Present: Using False Memory to Better Understand False Hearing in Young and Older Adults," *Memory & Cognition* 48 (2020): 1403–1416.

9. U. Lindenberger and U. Mayr, "Cognitive Aging: Is There a Dark Side to Environmental Support?," *Trends in Cognitive Sciences* 18 (2014): 7–15.

10. L. L. Jacoby, Y. Shimizu, K. Velanova, and M. G. Rhodes, "Age Differences in Depth of Retrieval: Memory for Foils," *Journal of Memory and Language* 52 (2005): 493–504.

11. J. S. McIntyre and F. I. M. Craik, "Age Differences in Item and Source Information," *Canadian Journal of Psychology* 41 (1987): 175–192.

12. G. O. Einstein and M. A. McDaniel, "Prospective Memory: Multiple Retrieval Processes," *Current Directions in Psychological Science* 14 (2005): 286–290.

13. J. D. Henry, M. S. MacLeod, L. H. Phillips, and J. R. Crawford, "A Meta-Analytic Review of Prospective Memory and Aging," *Psychology and Aging* 19 (2004) 27–39.

Chapter 7

1. K. S. Lashley, "Integrative Functions of the Cerebral Cortex," *Physiological Review* 13 (1933): 1–42.

2. W. Penfield and P. Perot, "The Brain's Record of Auditory and Visual Experience. A Final Summary and Discussion," *Brain* 86 (1963): 595–696.

3. For fuller details of this Nobel Prize–winning work, see M.-B. Moser, D. C. Rowland, and E. I. Moser, "Place Cells, Grid Cells, and Memory," *Cold Springs Harbor Perspectives in Biology* 7 (2015): 1–15.

4. E. A. Maguire, D. C. Gadian, I. S. Johnsrude, C. D. Good, J. Ashburner, R. S. J. Frackowiak, and C. D. Frith, "Navigation-Related Structural Changes in the Hippocampi of Taxi Drivers," *Proceedings of the National Academy of Sciences of U.S.A.* 97 (2000): 4398–4403.

5. B. Bowles, C. Crupi, S. M. Mirsattari, S. E. Pigott, A. G. Parrent, J. C. Pruessner, A. P. Yonelinas, and S. Köhler, "Impaired Familiarity with Preserved Recollection after Anterior Temporal-Lobe Resection That Spares the Hippocampus," *Proceedings of the National Academy of Sciences of U.S.A.* 104 (2007): 382–387.

6. F. I. M. Craik, M. B. Barense, C. J. Rathbone, J. E. Grusec, D. T. Stuss, F. Gao, C. J. M. Scott, and S. E. Black, "VL: A Further Case of Erroneous Recollection," *Neuropsychologia* 56 (2014): 367–380.

7. S. Kapur, F. I. M. Craik, E. Tulving, A. A. Wilson, S. Houle, and G. M. Brown, "Neuroanatomical Correlates of Encoding in Episodic Memory: Levels of Processing Effect," *Proceedings of the National Academy of Sciences of U.S.A.* 91 (1994): 2008–2011.

8. E. Tulving, S. Kapur, H. J. Markowitsch, F. I. M. Craik, R. Habib, and S. Houle, "Neuroanatomical Correlates of Retrieval in Episodic Memory:

Auditory Sentence Recognition," *Proceedings of the National Academy of Sciences of U.S.A.* 91 (1994): 2012–2015.

9. S. E. Petersen, P. T. Fox, M. I. Posner, M. Mintun, and M. E. Raichle, "Positron Emission Tomography Studies of the Cortical Anatomy of Single-Word Processing," *Nature* 331 (1988): 585–589.

10. E. Tulving, S. Kapur, F. I. M. Craik, M. Moscovitch, and S. Houle, "Hemispheric Encoding/Retrieval Asymmetry in Episodic Memory: Positron Emission Tomography Findings," *Proceedings of the National Academy of Sciences of U.S.A.* 91 (1994): 2016–2020.

11. B. P. Staresina and M. Wimber, "A Neural Chronometry of Memory Recall," *Trends in Cognitive Sciences* 23 (2019): 1071–1085.

12. Staresina and Wimber, "A Neural Chronometry of Memory Recall."

Chapter 8

1. A. R. Luria, *The Mind of a Mnemonist* (New York: Basic Books, 1968).

2. E. S. Parker, L. Cahill, and J. L. McGaugh, "A Case of Unusual Autobiographical Remembering," *Neurocase* 12 (2006): 35–49.

3. A. D. de Groot, *Thought and Choice in Chess* (The Hague: Mouton, 1965).

4. E. A. Maguire, E. R. Valentine, J. M. Wilding, and N. Kapur, "Routes to Remembering: The Brains behind Superior Memory," *Nature Neuroscience* 6 (2003): 90–95.

5. M. P. Walker and R. Stickgold, "Sleep, Memory, and Plasticity," *Annual Review of Psychology* 57 (2006): 139–166.

6. S. Colcombe and A. F. Kramer, "Fitness Effects on the Cognitive Function of Older Adults: A Meta-Analytic Study," *Psychological Science* 14 (2003): 125–130.

7. M. Husain and M. A. Mehta, "Cognitive Enhancement by Drugs in Health and Disease," *Trends in Cognitive Sciences* 15 (2011): 28–36.

8. K. Ball, D. B. Berch, K. F. Helmers, J. B. Jobe, M. D. Leveck, M. Marsiske, J. N. Morris, et al., "Effects of Cognitive Training Interventions with Older Adults: A Randomized Controlled Trial," *Journal of the American Medical Association* 288 (2002): 2271–2281.

9. J. M. Jennings and L. L. Jacoby, "Improving Memory in Older Adults: Training Recollection," *Neuropsychological Rehabilitation* 13 (2003): 417–440.

10. N. D. Anderson, P. L. Ebert, C. L. Grady, and J. M. Jennings, "Repetition Lag Training Eliminates Age-Related Recollection Deficits (and Gains Are Maintained after Three Months) but Does Not Transfer: Implications for the Fractionation of Recollection," *Psychology and Aging* 33 (2018): 93–108.

11. D. C. Park, J. Lodi-Smith, L. Drew, S. Haber, A. Hebrank, G. N. Bischof, and W. Aamodt, "The Impact of Sustained Engagement on Cognitive Function in Older Adults: The Synapse Project," *Psychological Science* 25 (2014): 103–112.

12. J. A. Anguera, J. Boccanfuso, J. L. Rintoul, O. Al-Hashimi, F. Faraji, J. Janowich, E. Kong, et al., "Video Game Training Enhances Cognitive Control in Older Adults," *Nature* 501 (2013): 97–101.

13. C. Lustig, P. Shah, R. Seidler, and P. A. Reuter-Lorenz, "Aging, Training, and the Brain: A Review and Future Directions," *Neuropsychology Review* 19 (2009): 504–522.

FURTHER READING

Baddeley, Alan, Michael Eysenck, and Michael Anderson. *Memory*. 3rd ed. New York: Psychology Press, 2020. A comprehensive introduction to the study of human memory and its applications by leading experts in the field.

Craik, Fergus. *Remembering: An Activity of Mind and Brain*. Oxford: Oxford University Press, 2021. A more extensive (and more technical) personal account of human memory as a type of processing by one of the present authors.

Draaisma, Douwe. *Metaphors of Memory: A History of Ideas about the Mind*. New York: Cambridge University Press, 2000. A fascinating survey of ideas on memory across the centuries.

Duhigg, Charles. *The Power of Habit*. New York: Random House, 2012. An exploration of the science behind habit creation and recreation by an award-winning former *New York Times* business reporter.

Einstein, Gilles, and Mark McDaniel. *Memory Fitness: A Guide for Successful Aging*. New Haven, CT: Yale University Press, 2004. A well-written account by two prominent researchers of how memory changes with age and ways to reduce the problems associated with aging.

Heath, Chip, and Dan Heath. *Made to Stick: Why Some Ideas Survive and Others Die*. New York: Random House, 2007. An entertaining account of some practical ways to improve memory.

Kandel, Eric. *In Search of Memory: The Emergence of a New Science of Mind*. New York: W. W. Norton, 2006. A personal history of the developing neurobiology of memory by a Nobel Prize winner.

Neisser, Ulric, and Ira Hyman Jr. *Memory Observed: Remembering in Natural Contexts*. New York: Worth Publishers, 1999. Selections, with commentaries, of interesting articles dealing with memory in natural surroundings.

Schacter, Daniel. *The Seven Sins of Memory: How the Mind Forgets and Remembers*. Boston: Houghton Mifflin, 2001. A leading researcher's well-written description of everyday memory problems and how to avoid them.

INDEX

Page numbers in italics refer to figures.

FERGUS CRAIK taught in the psychology department at the University of Toronto and then worked at the Rotman Research Institute in Toronto until his retirement. He is a foreign honorary member of the American Academy of Arts and Sciences.

LARRY JACOBY was a faculty member in the psychology department at Washington University in Saint Louis. He is a member of the American Academy of Arts and Sciences.